Cork Science and its Applications
CSA2017

Edited by

Ricardo Sousa
Ofélia Anjos

Cork and cork agglomerates are natural cellular materials that are receiving much interest in a number of technological applications, e.g. as Feedstock for the Additive 3D Printing Technology; Solar, Wind and Bioenergy applications; Sorbent for Pesticides and Heavy Metals; and Water Treatment. Experimental, analytical and computational research is reported, as well as new applications in fields ranging from design and architecture to mechanical, chemical, civil and electronics engineering.

Cork Science and its Applications
CSA2017

1st International Conference on Cork Science and its Applications (CSA2017), Aveiro, Portugal, 7-8 Sep., 2017

Ricardo Sousa[1] and Ofélia Anjos[2]

[1]Department of Mechanical Engineering, University of Aveiro, Portugal

[2]Agrarian School of Polytechnic Institute of Castelo Branco, Castelo Branco, Portugal

Peer review statement

All papers published in this volume of "Materials Research Proceedings" have been peer reviewed. The process of peer review was initiated and overseen by the above proceedings editors. All reviews were conducted by expert referees in accordance to Materials Research Forum LLC high standards.

Published as part of the proceedings series
Materials Research Proceedings
Volume 3 (2017)

ISSN 2474-3941 (Print)
ISSN 2474-395X (Online)

ISBN 978-1-94529140-1 (Print)
ISBN 978-1-94529141-8 (eBook)

This book contains information obtained from authentic and highly regarded sources. Reasonable efforts have been made to publish reliable data and information, but the author and publisher cannot assume responsibility for the validity of all materials or the consequences of their use. The authors and publishers have attempted to trace the copyright holders of all material reproduced in this publication and apologize to copyright holders if permission to publish in this form has not been obtained. If any copyright material has not been acknowledged please write and let us know so we may rectify in any future reprint.

Distributed worldwide by

Materials Research Forum LLC
105 Springdale Lane
Millersville, PA 17551
USA
http://www.mrforum.com

Manufactured in the United States of America
10 9 8 7 6 5 4 3 2 1

Table of Contents

Preface

The chapters collected in this book edition were presented at the first international conference on cork science and its applications (CSA2017), held in the beautiful city of Aveiro, Portugal, September 7th-8th, 2017. This conference aimes to become a forum for discussions between young and senior scientists, academia and industry, generating contacts and bringing up new ideas and technologies in the area of cork material and its applications. In an effort to better understand the potential and various features of this natural cellular material, many research studies have been carried out using experimental, analytical and computational methods. The great advances in material development and modelling have been accompanied by the expansion in the range of applications, covering now not only design and architecture, but also aspects of mechanical, chemical, civil and electronics engineering. Moreover, considerable progress in the area of computational tools has enabled advanced predictive modelling of material behavior under different loading conditions.

We deeply acknowledge the support given by Amorim Cork Composites (ACC) who helped organize this conference. A word of gratitude also to the University of Aveiro, particularly de Department of Mechanical Engineering and the Center of Mechanical Technology and Automation (TEMA) and to the Portuguese Cork Association (APCOR), always pushing forward the innovation on cork material. Finally, we would like to take the opportunity to give a special thanks to all the members of the scientific committee and invited reviewers for their invaluable contribution to assuring the relevance and quality of submitted contributions.

We are looking forward to seeing you at the next CSA conference, in Barcelona!

Ricardo Sousa and Ofélia Anjos (editors)

Committees

Scientific Committee:

- Ana Cláudia Dias (U.Aveiro, DAO, PT)
- Ana Cristina Esteves (U.Aveiro, DBio, PT)
- António Bastos Pereira (U. Aveiro, DEM, PT)
- Armando Silvestre (U.Aveiro, DQ, PT)
- Emanuel Fernandes (U.Minho, PT)
- Fábio Fernandes (U.Aveiro, DEM, PT)
- Filipe Teixeira-Dias (U.Edinburgh, UK)
- Francisco G. Rosales (U.Porto, FEUP, PT)
- Helena Pereira (U. Lisboa, PT)
- José Martinho Oliveira (U.Aveiro, ESAN, PT)
- José Pedro Sousa (U.Porto, FAUP, PT)
- Maria Verdum (I.C. Suro, ES)
- Mariusz Ptak (P. Wroclawska, PL)
- Ofélia Anjos (I.P.Castelo Branco, PT)
- Paula Marques (U.Aveiro, DEM, PT)
- Paulo Vila Real (U.Aveiro, DECivil, PT)
- Raphael Duval (Naturality, FR)
- Ricardo Alves de Sousa (U.Aveiro, PT), conference chair
- Susana Silva (Amorim Cork, ACC, PT)
- Teresa Franqueira U.Aveiro, DECA, PT)

Cork Science and its Applications (CSA2017)   Materials Research Forum LLC
Materials Research Proceedings 3 (2017) 1-10   doi: http://dx.doi.org/10.21741/9781945291418-1

Thermal Conductivity of Agglomerate Cork

Tiago Santos[1,2,a*], João S. Amaral[1,b], Vítor A. F. Costa[2,c], Vítor S. Amaral[1,d]

[1]University of Aveiro, CICECO – Aveiro Institute of Materials, Department of Physics, Campus Santiago, 3810-193 Aveiro, Portugal

[2]University of Aveiro, Center for Mechanical Technology and Automation, Department of Mechanical Engineering, Campus Santiago, 3810-193 Aveiro, Portugal

[a]tiago.santos@ua.pt, [b]jamaral@ua.pt, [c]v.costa@ua.pt, [d]vamaral@ua.pt

Keywords: Agglomerate Cork, Thermal Conductivity, Resin, Compacting Pressure

Abstract. For fine grains, agglomerate cork presents essentially isotropic properties, which depend mainly on the grain size, resin type, resin percentage and compacting pressure. Cork is a well-known material due to its very interesting mechanical properties and low thermal conductivity. Mechanical properties of agglomerate cork strongly depend on the referred parameters, and additional work needs to be conducted to evaluate how they affect its thermal properties. Experimental work to measure the thermal conductivity and specific heat of agglomerate cork is described, and the obtained results analysed. Samples, in the form of cubes of agglomerate cork, are experimentally tested to evaluate the agglomerate cork thermal conductivity and specific heat. Different cork grain sizes, resin types, resin percentages, and compacting pressures are considered. Thermal conductivity and specific heat are measured using the Hotdisk[©] technique. Experimental results of the thermal conductivity and specific heat of the agglomerate cork are complemented by evaluating the influence of agglomerate grain size, nature of the used resin, percentage of resin used, and the compacting pressure imposed during the resin curing and consolidation process. It is concluded that the agglomerate cork thermal properties are only slightly affected by the referred parameters, the agglomerate cork density, related with the compacting pressure, having the highest influence over those thermal properties.

Introduction

Cork is an increasingly important material, with more and more practical applications, due to its very interesting and important mechanical and thermal properties [1]. Cork can be used for many purposes in its natural state, cork parts being extracted from bark, or as agglomerate cork. In this case a resin is mixed with the cork granules, which can have different sizes, the mixture is deposited in a mould container where it is subjected to some compressive compacting pressure, and the consolidate part of agglomerate cork is obtained after the resin cure. Some time is needed for resin cure and consolidation of the agglomerate cork, which can be shortened increasing the temperature of the cure and consolidation process, but in a limited way as to not start cork damage. Properties of natural cork exhibit some anisotropic behaviour, due to the different orientations acquired by bark during the cork tree growth. The agglomerate cork, instead, exhibits an essentially isotropic behaviour for all its properties, isotropy increasing as the grain size decreases and density increases.

Mechanical properties of agglomerate cork are severely affected by grain size, nature and percentage of resin, and compacting pressure imposed during the fabrication of the agglomerate cork [2]. Thus, different mechanical performances of agglomerate cork can be tuned selecting the adequate different grain sizes, different compositions and different compacting pressures. It is to be noted that different compacting pressures have a major influence on the density of the resulting agglomerate cork. The most relevant mechanical properties of the agglomerate cork include impact resistance and dampening, vibration absorption and dampening, and sound absorption and attenuation. The most relevant thermal property of the agglomerate cork is its thermal conductivity.

The main question addressed in this work is to evaluate how the thermal properties of the agglomerate cork, and especially its thermal conductivity and specific heat, are affected by the aforementioned parameters. It is also to be evaluated if this dependence needs to be taken into consideration for practical purposes, and for thermal design. This work relies mainly on the experimental measurements of the thermal conductivity and specific heat of agglomerate cork, for different grain sizes, different type and percentage of resin, and different compacting pressures.

Other thermal properties of the agglomerate cork can be considered and studied, including thermal diffusivity, even if it can be obtained from thermal conductivity, density and specific heat $\left[\alpha = k/(\rho c_P) \right]$, where c_P is the unit mass based constant pressure specific heat. However, the use of agglomerate cork for thermal purposes usually relies mainly on its low thermal conductivity, with advantageous performance as a thermal insulator. In many practical applications, mechanical and thermal advantages of the agglomerate cork use can be simultaneously accomplished.

Materials and methods
Different samples of agglomerate cork are considered, in the form of cubes with 40 mm side, the design of experiments being summarized in Table 1. Cork grains were obtained from a Portuguese granulate cork manufacturer, and they can be taken as representative for cork in general. Grain sizes are classified into two groups: Small, if the grain size is between 0.5 and 1.0 mm, and Large if the grain size is between 2.0 and 4.0 mm. Fig. 1 presents pictures of the samples with Small and Large grain sizes, where that difference is clear.

Fig. 1 Samples pictures of agglomerate cork of Small (left) and Large (right) grain sizes.

Cork Science and its Applications (CSA2017)   Materials Research Forum LLC
Materials Research Proceedings 3 (2017) 1-10   doi: http://dx.doi.org/10.21741/9781945291418-1

The resines employed in samples preparation were obtained from Flexpur, a Portuguese resin manufacturer, which are usually used for agglomerate cork manufacture. Only a qualitative classification was considered for the resin type, which can be flexible [Methylene Diphenyl DiIsocyanate (MDI) based], intermediate or hard [Tolunene DiIsocyanate (TDI) based] once cured. Resin content is quantified in a percentage weight basis, and it is a low value but high enough to lead to consolidate agglomerate cork.

Measurements of the thermal conductivity and specific heat are performed using the Gustafsson Probe Method (or Hot Disk) with the Thermal Constant Analyzer TPS 2500S. This transient method uses an electrically conducting pattern (Nickel) element acting both as a temperature sensor and heat source, insulated with two thin layers of Kapton (70 μm). The TPS element is typically assembled between two samples of similar characteristics with both faces in contact with the sensor surface. Measurement range is typically from 0.005 up to 1500 W/m.K. Sensor elements radius available range typically from 0.5 to 29.5 mm. For isotropic samples, as it is essentially the present case of the agglomerate cork, the Hot Disk method allows the determination of the thermal conductivity, thermal diffusivity and specific heat. For anisotropic samples axial/radial components can also be measured using adapted methods/software. The Hot Disk method is an international standard for measuring thermal conductivity and thermal diffusivity [3,4], with the ISO 22007-2 designation.

Table 1 Sample parameters.

Resin type	Resin content [wt.%]	Density [kg/m^3]	Grain Size [mm]
Flexible (1)	15	200	Small (0.5-1.0)
Intermediate (2-3)	15	200	Small (0.5-1.0)
Hard (4)	15	200	Small (0.5-1.0)
Flexible (1)	5	200	Small (0.5-1.0)
Flexible (1)	10	200	Small (0.5-1.0)
Flexible (1)	10	120	Small (0.5-1.0)
Flexible (1)	10	160	Small (0.5-1.0)
Flexible (1)	10	200	Large (2.0-4.0)

The (volumetric) specific heat of the samples can be determined independently with the Hot-Disk TPS 2005S, using a special sensor-cell. This cell consists of a heated gold cup coupled to a Hot Disk sensor in which the sample is placed. Two measurements of the temperature of the cell (empty and with the sample) under electric heating are required. The measurement range of the volumetric specific heat is from 0.005 up to 5 MJ/m^{3}K.

Unless differently specified, all measurements were obtained near room temperature (around 20°C).

Results and Analysis

Obtained experimental results for thermal conductivity and volumetric specific heat are summarized in Table 2. The reported uncertainty arises from the analysis of 10 independent measurements, for each data point listed.

Measured values of the agglomerate cork thermal conductivity are between 0.05 W/mK and 0.06 W/mK, which are of the same order of magnitude as those referred in the literature for agglomerate cork [5,6]. It is also observed that these values are close to the ones referred in the literature for natural cork [1,7]. Cork is made of cells enclosing air, where heat transfer through cork occurs mainly as heat conduction through the cells walls and also through the air enclosed in the cells, as cells sizes are too small for convection to occur. As the volume fraction of solid material (cells walls) is small, the overall cork material thermal conductivity is mainly conditioned by the thermal conductivity of the enclosed air, for both the natural cork and for the agglomerate cork.

Table 2 Summary of results

Temperature [°C]	Resin %	Resin type	Density [kg/m³]	Grain size	k [W/mk]	Uncertainty	c_P [MJ/m3K]	Uncertainty
21.59	10	1	200	2	0.05761	3.1E-04	0.36851	0.0051
18.76	10	1	120	1	0.05000	1.5E-04	0.24909	0.0056
19.48	10	1	160	1	0.05276	2.2E-04	0.30113	0.0069
19.90	10	1	200	1	0.05935	3.2E-04	0.42638	0.0097
24.26	5	1	200	1	0.05702	3.4E-04	0.34751	0.0097
20.13	15	1	200	1	0.05631	2.6E-04	0.35801	0.0060
19.97	15	2	200	1	0.05568	1.5E-04	0.35684	0.0054
18.71	15	3	200	1	0.05726	2.2E-04	0.37686	0.0120
20.64	15	4	200	1	0.05672	1.5E-04	0.34401	0.0030

Even if with a small number of measurements, Fig. 2 shows essentially a slight decrease on the thermal conductivity and on the volumetric specific heat of agglomerate cork as the grain size increases. Even so, these changes on the thermal properties of the agglomerate cork, solely induced by grain size differences, are only residual, with little relevance on the thermal performance of the agglomerate cork in practical use.

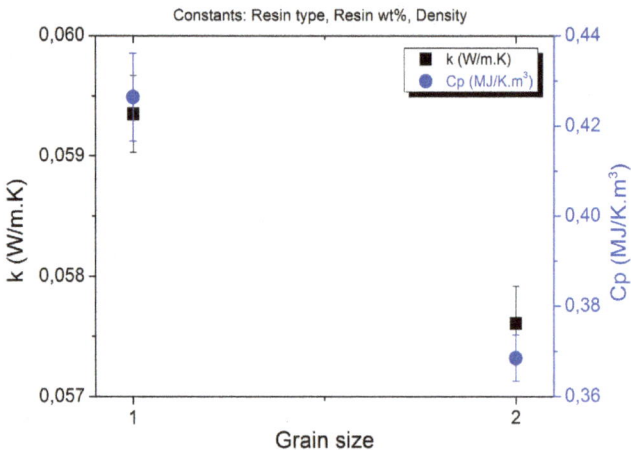

Fig. 2 Influence of the grain size on the thermal conductivity and volumetric specific heat of the agglomerate cork.

Cork Science and its Applications (CSA2017)   Materials Research Forum LLC
Materials Research Proceedings **3** (2017) 1-10   doi: http://dx.doi.org/10.21741/9781945291418-1

Fig. 3 shows the dependence of the thermal conductivity and volumetric specific heat of the agglomerate cork on the different types of resins used. From the measured values it is difficult to find a pattern relating the type of resin with the agglomerate cork thermal conductivity and volumetric specific heat, even if a maximum of these two thermal properties seems to exist for the resin type 3. However, this change on the thermal conductivity and specific heat of the agglomerate cork induced by the different types of resin is only residual, and with small relevance only on the thermal performance of the agglomerate cork in practical use.

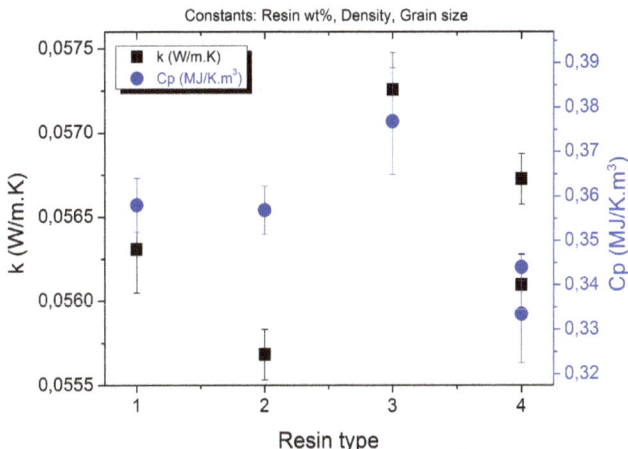

Fig. 3 Influence of the resin type on the thermal conductivity and volumetric specific heat of the agglomerate cork.

Fig. 4 shows the dependence of the thermal conductivity and volumetric specific heat of the agglomerate cork induced by the percentage (in weight) of the resin used. It is observed that the agglomerate cork thermal conductivity and specific heat increase as increases the percentage of resin for small percentages of resin, and that for resin percentages higher than 10% this behaviour is inverted for both thermal properties. Even so, and as for the previous cases where other different parameters were changed, this change on the thermal conductivity and volumetric specific heat of the agglomerate cork induced by the different percentage of resin is only residual, and with small relevance for the thermal performance of the agglomerate cork in practical use.

For all the measurements made both thermal conductivity and volumetric specific heat of the agglomerate cork present essentially the same variations with changes on the considered parameters. This similar behaviour is observed in both the sense (increase or decrease) of the changes, and in the relative change of the numerical values of the thermal properties under analysis.

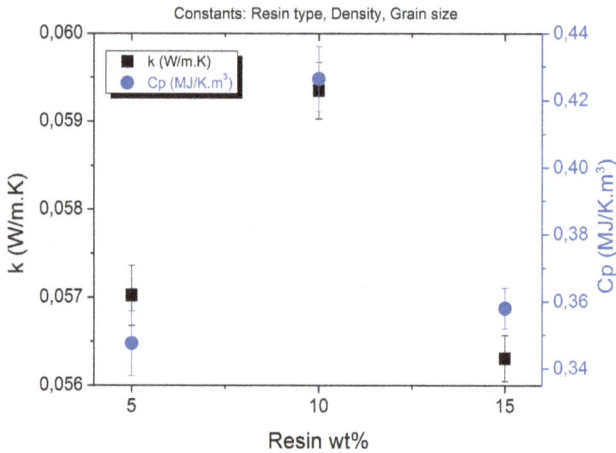

Fig. 4 Influence of the resin percentage (in weight) on the thermal conductivity and volumetric specific heat of the agglomerate cork.

Fig. 5 shows pictures of agglomerate cork samples with Small grain size for different compacting pressures, where it is clear a denser medium for higher compacting pressures.

Fig. 5 Sample pictures of agglomerate cork with Small grain size for lower (left) and higher (right) compacting pressures.

Fig. 6 shows the dependence of the thermal conductivity and volumetric specific heat of agglomerate cork on its density. Higher densities of agglomerate cork are obtained for smaller grain sizes and/or for higher compacting pressures. Higher densities correspond to higher volume fractions of the solid materials (cells of small dimensions) and thus to lower volume fractions of the air enclosed in the cells, thus leading to higher thermal conductivities and to higher specific heats. It is thus observed the increase on the agglomerate cork thermal conductivity and specific heat as its density increases. Even if the influence of the density of the agglomerate cork on its thermal conductivity is small, it is higher than for the preceding analysed parameters (grain size, resin type, and resin percentage). Volumetric specific heat is a thermal property associated with the mass contained in a given volume, denser media leading to higher specific heats. Even so, for moderate compacting pressures, the thermal conductivity and specific heat only slightly change (increase) with the compacting pressure. It is to be noted that density can be increased by increasing the compacting pressure when obtaining the agglomerate cork, or when in use the agglomerate cork is subjected to compressive loads. Care needs to be taken in what concerns the dependence of the agglomerate cork thermal properties on density, as it can be used in service under different compressive loads, which can be considerably high and changing with time.

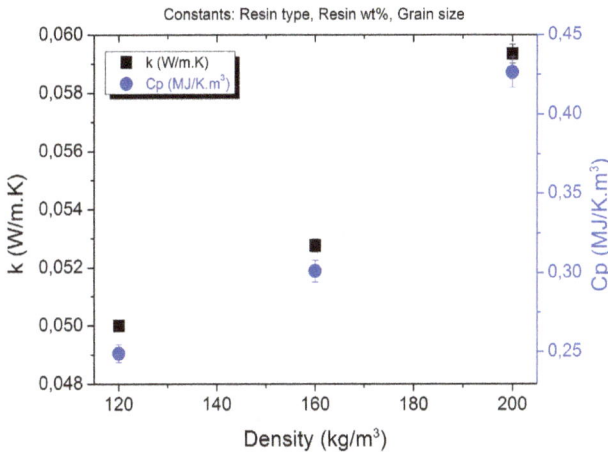

Fig. 6 Influence of the agglomerate density on its thermal conductivity and volumetric specific heat.

It is also to be observed from Table 2 the small uncertainty on the measured thermal conductivities and volumetric specific heats, which are associated to a relatively high number of measurements (10) for each sample, thus indicating a small variance on the numerical values of these thermal properties.

Comparing the measured thermal conductivity with experimental data from the literature, Fig. 7 shows that very good agreement exists, both in what concerns the numerical values of the thermal conductivity and on how it is affected by the agglomerate cork density. This result reinforces the previous presented argument, since heat conduction in cork and cork agglomerates will be mainly influenced by the air volume enclosed within the cork cells. Lowering the available air volume by cell compression (increasing density) will lead to an increase in thermal conductivity. This increase is here shown to follow an approximately linear relation of thermal conductivity with density. By extrapolating the observed linear behaviour down to low values of density, a fair comparison with the value of the thermal conductivity of air at ambient pressure and temperature is obtained. Interestingly, by extrapolation the observed linear relation to higher values of density, namely to 1200 kg/m^3 , the estimated density of cork cell walls, from their chemical formula [8], one obtains a thermal conductivity value of 0.19 W/m.K in very good agreement with the reported value of 0.2 W/m.K for the thermal conductivity values of cork cell walls [7]. This linear relationship between the density and thermal conductivity of cork and cork agglomerates is here shown to hold between the low and high density end points.

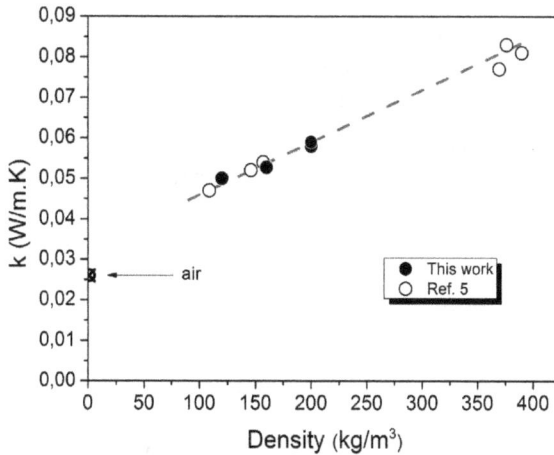

Fig. 7 Comparison of the measured thermal conductivity with data from Ref. 5.

Temperature is another parameter influencing the thermal conductivity and specific heat of the agglomerate cork. Table 3 contains the thermal conductivity and specific heat of two different samples of agglomerate cork (of Grain size 1) for different temperatures showing that higher temperatures lead to slightly higher thermal conductivities and volumetric specific heats, even if only with small influence on the thermal performance of the agglomerate cork when in use. It is to be noted that the considered different temperatures are closer to the usual room temperature. Care needs to be taken in what concerns the dependence of the agglomerate cork thermal properties on temperature, as the agglomerate cork can be used in service at temperatures, which can even change with time, considerably different from the usual room temperature with somewhat different thermal properties than expected at the room temperature.

Table 3 Influence of the sample's temperature on the measured thermal properties.

Resin %	Resin type	Density [kg/m³]	Grain size	Temperature [°C]	N° of measures	k [W/m.K]	Uncertainty	c_p [MJ/m³K]	Uncertainty
15	1	200	1	21.40	3	0.05610	2.4E-04	0.34233	0.0014
				24.26	12	0.05702	3.4E-04	0.34751	0.0097
15	3	200	1	18.71	12	0.05726	2.2E-04	0.37686	0.0120
				22.64	4	0.05882	1.7E-04	0.38923	0.0650
15	4	200	1	19.38	10	0.05610	1.8E-04	0.33346	0.0110
				20.64		0.05672	1.5E-04	0.34401	0.0030

Conclusions

From the measurements and analyses made in this work it is concluded that the considered parameters (grain size, resin type, resin percentage, agglomerate density, and temperature) influence the thermal conductivity and specific heat of the agglomerate cork. Even so, and considering the range of parameters considered in this experimental study, the main conclusion is that the thermal properties of the agglomerate cork are not considerably affected by those parameters. Among the considered parameters, the agglomerate cork density shows the higher, but still an overall small, effect on the thermal properties of the agglomerate cork. Compacting pressure is associated with the agglomerate cork manufacture, but the compressive loads to which the agglomerate cork is subjected when in service is dependent of the particular application conditions. Thus, density of agglomerate cork when in use needs to be regarded not only as a material property but also depending on the particular conditions of use, which affect (increase) the agglomerate cork thermal properties.

The considered parameters considerably change the mechanical properties of the agglomerate cork, and their control/change can be used for tuning/adjust its expected mechanical properties. However, for contained variations of those parameters, the agglomerate cork thermal properties are only slightly affected by those parameters, with marginal relevance only on the thermal performance for the main usual applications of the agglomerate cork. Thus, the dependence of the agglomerate cork thermal properties on the referred parameters can be neglected for many thermal designs. However, for high compressive loads and/or for temperatures far from the usual room temperature influence of density and temperature on the agglomerate cork thermal properties cannot be neglected.

Acknowledgements

This work was developed within the scope of the project CICECO-Aveiro Institute of Materials, POCI-01-0145-FEDER-007679 (FCT Ref. UID /CTM /50011/2013), financed by national funds through the FCT/MEC and when appropriate co-financed by FEDER under the PT2020 Partnership Agreement. J. S. Amaral acknowledges FCT research grant IF/01089/2015.

References

[1] H Pereira, Cork: Biology, production and uses, Elsevier, Amsterdam, 2007.

[2] P.T. Santos, S. Pinto, P.A.A.P.M.Marques, A.B. Pereira, R.J. Alves de Sousa, Agglomerated cork: how to tailor its mechanical properties, Composite Structures (Elsevier), submitted, 2017. https://doi.org/10.1016/j.compstruct.2017.07.035

[3] S. E. Gustafsson, Transient plane source technique for thermal conductivity and thermal diffusivity measurements of solid materials, Revue of Scientific Instruments, Vol. 62, No. 3, pp. 797-804, 1991. https://doi.org/10.1063/1.1142087

[4] M. Gustavsson, E. Karawacki, S. E. Gustafsson, Thermal conductivity, thermal diffusivity, and specific heat of thin samples from transient measurements with hot disk sensors, Revue of Scientific Instruments, Vol. 65, No. 12, pp. 3856-3859, 1994. https://doi.org/10.1063/1.1145178

[5] F. Barreca, C. R Fichera, Thermal insulation characteristics of cork agglomerate panels in sustainable food buildings, Proc. Of the 7th International Conference on Information and Communication Technologies in Agriculture, Food and Environment (HAICTA 2015) Kayla, Greece, 17-20 September 2015.

[6] F. Barreca, C. R Fichera, Thermal insulation performance assessment of agglomerate cork boards, Wood and Fiber Science, 48 (2) 2016 pp. 1-8.

Cork Science and its Applications (CSA2017)   Materials Research Forum LLC
Materials Research Proceedings **3** (2017) 1-10   doi: http://dx.doi.org/10.21741/9781945291418-1

[7] S. P .Silva, M. A, Sabino, E. M. Fernandes, V. M. Correlo, L. F. Boesel, R. L. Reis, Cork: Properties, capabilities and applications. International Materials Reviews, 50 (6) 2006 pp. 345-365. https://doi.org/10.1179/174328005X41168

[8] L. J. Gibson, K. E. Easterling and M. F. Ashby, The structure and mechanics of cork, Proceedings of the Royal Society of London A: Mathematical, Physical and Engineering Sciences, 1981, A377, 99–117. https://doi.org/10.1098/rspa.1981.0117

Cork Science and its Applications (CSA2017)
Materials Research Proceedings **3** (2017) 11-18

Materials Research Forum LLC
doi: http://dx.doi.org/10.21741/9781945291418-2

Revaluation of Catalan Low Quality Cork as Feedstock for the Additive 3D Printing Technology

M. Verdum[1a*], C. Sánchez[2b], J. Castro[1c], J. Lladó[1d] and P. Jové[1e]

[1] Catalan Cork Institute, R+D Department, Miquel Vincke i Meyer, 13 – 17200 Palafrugell, Spain

[2] 3D Spider Print, Pol. Ind . La Bòvila Nau1 - 08232 Viladecavalls, Spain

[a]mverdum@icsuro.com, [b]csanchez@3dspiderprint.com, [c]jcastro@icsuro.com,
[d]jllado@icsuro.com, [e]pjove@icsuro.com

Keywords: Cork, Coil, 3D Printing, Plastic Biopolymer

Abstract. Recently 3D printing polymeric coils have been introduced to the market combined by a plastic base and vegetal support. These dual coils own the features of biopolymer and physicochemical characteristics of the plant which it has been mixed with, resulting in a new material with a high added value. Cork properties are able to provide versatility in terms of 3D printing technology and directly compete with other plastic-based products, which have a greater environmental impact. Low quality cork is not suitable for the manufacturing of cork stoppers: byproducts, first harvesting cork, dust or even burnt cork. The use of low quality cork will increase the economic value of cork encouraging forest owners to manage their forests, which are unmanaged in Catalonia at 50%. The new material will be ecosostenible because it is PLA based and the cork percentage is greater than 20%. This study compares the physical, chemical and mechanical properties of pellets, the studied cork coil and the commercial cork coil. In general, pellets and the studied cork coil showed a higher percentage of cork than the commercial cork coil. The presence of cork material in a 3D filament can be evaluated using a chemical composition assays and scanning electron microscopy. The percentage of suberin is related to the filament cork content and allows us to compare different commercial products.

Introduction

According to Allied Market Research, 3D printing is referred as additive printing technology that enables manufacturers to develop objects using a digital file and variety of printing materials. The global market for 3D printing material includes polymers, metals and ceramics. In addition, 3D printing offers a wide array of applications in various industries, namely consumer products, industrial products, defense and aerospace, automotive, healthcare, education and research.

The plastic coils have been combined with wood, coconut fiber or carbon named as dual coils. In these three cases polylactic acid (PLA) has been used as a binding polymer. Those interesting materials have different properties according to the natural material added to and its proportion. These dual coils own the versatility of the biopolymer and the physicochemical characteristics of the plant which it has been mixed with, causing a new material with a high added value. For example, coconut and wood coil are a mixture of 40% coconut and wood particles respectively crushed together by the plastic polymer. The prints actually look and smell like wood and post-processing presents various options, as the prints can be sanded, grinded and painted like standard wooden products. In the case of carbon fiber coils have 15% of this compound and have been used in the automation industry to make less heavy prototypes.

Cork Science and its Applications (CSA2017)   Materials Research Forum LLC
Materials Research Proceedings 3 (2017) 11-18   doi: http://dx.doi.org/10.21741/9781945291418-2

3D Printing Market Report, published by Allied Market Research, forecasts that the global market is expected to garner \$8.6 billion by 2020, registering a CAGR of 21% during the period 2015-2020. This surge in growth is primarily attributed to the rising demand for faster and efficient ways to manufacture complex design objects using a wide array of materials.

Cork is a natural material, renewable and biodegradable with a combination of properties that make it unique and versatile. These properties include its low density, high mechanical strength and fire, low thermal and electrical conductivity, as well as being a good thermal insulation, acoustic insulation and possess a high elasticity. These properties can provide a lot of versatility and possibilities in terms of 3D printing technology and directly compete with other plastic-based products, which have a greater environmental impact. Low quality cork is not suitable for manufacturing cork stoppers: byproducts, first harvesting cork, dust or even burnt cork. The use of low quality cork will increase the economic value of cork encouraging forest owners to manage their forests, which are unmanaged in Catalonia at 50%.

3D Spider Print and Catalan Cork Institute are working to find a new duel cork-coin. The study aims to create a new product with cork base additive usable for printing, extrusion, injection and other manufacturing processes. This product corresponds to a mixture of granulated cork and plastic biopolymer (as used in 3D printing). The new material will be ecosostenible because of being PLA based and having a cork percentage greater than 20%.

3D filament manufactures are looking for innovative means expanding sales into the industrial prototyping and additive manufacturing market [1]. 3D printing is a new technology for making objects by building up layers of a given material, usually plastic or metal. The material is fed into 3D printers in the form of a filament, which is heated so that it liquefies and then solidifies one layer at a time. Cork oak forests have to be managed for about 60 years before they produce high-quality cork. During this time large amounts of low quality cork is produced, and there is little demand for it as a product. In Catalonia, lacking of economic incentive means forests are left unmanaged and this has led to large forest fires. These fires left behind huge amounts of burnt cork oak groves [2]. Catalan Cork Institute research had shown that the burnt cork can be used in new applications, such as 3D printing, therefore, offering a new opportunity for forest managers to still make a profit from these areas and to manage them again in following years. The plastic coils have been combined with wood, coconut fiber, carbon and even cork. Thus, the results exposes above are a comparative between cork, commercial filament and a new dual cork coil (studied cork coil).

The aim of this study is to reevaluate Catalan low-quality cork, introducing it to the 3D technology and thus giving it a new application.

Material and methods
Samples
Cork granules were prepared by cutting planks of natural cork from the North East of Spain (Catalonia). The cork particle size lower than 250 µm is considered cork dust, thus cork dust is a by-product [3]. In this study, we used cork dust generated during the preparation and cutting cork planks intended for sparkling wine. The chemical mix with PLA and cork dust was studied by 3DSpider Print. A dual bis machine was needed due to the low density of granulate cork. A sieve has been used to strain cork dust at 250µm. Then cork dust lower than 250 µm particle size were introduced in a turbo-mixer and acid polilactid was added too. After this process, pellets were made that can be stretched to obtain coils (filaments).

Polylactic acid or polylactide (PLA) is a biodegradable and bioactive thermoplastic aliphatic polyester derived from renewable resources, such as corn starch, tapioca roots, chips or starch or

Cork Science and its Applications (CSA2017)   Materials Research Forum LLC
Materials Research Proceedings 3 (2017) 11-18   doi: http://dx.doi.org/10.21741/9781945291418-2

sugarcane. Pellets and the studied cork coil are lightweight cork-filled PLA-based filament which are gravimetrically filled with approximately 25% cork dust. Different proportions of cork dust (10-20-30%) were tested. This study compares the physical, chemical and mechanical properties of pellets, the studied cork coil and a commercial cork coil (Easy Cork of Form Future).

Chemical & physico-mechanical characterization
The summative chemical analyses included the determination of extractives, suberin, lignin and holocellulose content. The studied pellets and cork filaments were cut by an Ultra centrifugal Mill ZM 200 at 0,75mm.

The methodology used is an adaptation of chemical composition of Jové P. et al. 2011 [4], and is described above. Extractives were removed by successive Soxhlet extractions with dichloromethane (6h), ethanol (8h) and hot water (17h). After each extraction step the solution was evaporated and the solid residue was weighed with an analytical balance. The suberin content was analyzed in extractive-free material by methanolysis for depolymeritzation. The desuberized fraction was used for subsequent analyses. Klason lignin or acid-insoluble lignin were determined by acid hydrolysis. The residue was washed with hot water, dried and boiling. The filtrate was used to determine acid soluble lignin by measuring the absorbance. Klason lignin and acid-soluble lignin were combined to give the total lignin content [5]. Holocellulose was isolated from the desuberized fraction by delignification for 2h using the acid chloride method [11].

The dual cork coil morphology and structure was revealed by scanning electron microscopy (SEM) and Fourier Transform Infrared Spectroscopy (FTIR). The SEM samples have been placed on a stub and evaporated carbon (Emitech, German, K950 turbo evaporator). Examinations were carried out with a scanning electron microscopy FE-SEM Hitachi, Japan, S-4100. Digital images was collected and processed by Quarz PCI program. The results were compared with the commercial cork filament.

The FTIR test was carried out using a Cary 630 FTIR equipment to study the vibration and rotation of the molecules in the infrared region of the electromagnetic spectrum.

Print characteristics
Print characteristics were also done. The filaments were tested with a Delta printer WASP 20 40 with software Cura 14.07.

Results and discussion
Chemical & physico-mechanical characterization
The chemical analysis that were used for the chemical composition evaluation is the methodology used for chemical composition of bark layers of *Quercus suber* L. Results for chemical analysis of the commercial coil, the pellets and the studied coil are shown in table 1.

Table 1 Chemical composition of commercial cork coil, pellets and studied cork coil.

	Commercial coil	Pellets	Studied coils	Cork [4]
DCM extractives	91,6%	19,2%	74,2%	
Et-OH extractives	0,3%	1%	0,8%	9-20%
H₂O extractives	1%	2,9%	2,4%	
Suberin content	3,6%	25,4%	5,8%	30-60%
Lignin and holocellulose content	3,5%	51,5%	16,8%	12-22%

These results were different than previously described cork samples [4]. Cork, the outer bark of *Quercus suber* L., is a plant tissue composed of suberin (30-60%), lignin (19-22%), polysaccharides (12-20%) and extractives (9-20%) [6].

The chemical composition of the commercial coil showed higher percentages of dichloromethane extractives than the pellets and the studied coil. The commercial coil also had a lower percentages of ethanol and water extractives than the pellets or the studied coil. This could be explained because PLA (the majority compound) being an aliphatic polyester. It seems that it should be solubilised with a low-polarity solvent, like dichloromethane. According to these results, the commercial coil would seem to contain more PLA than the studied coil (Fig.1 A and B).

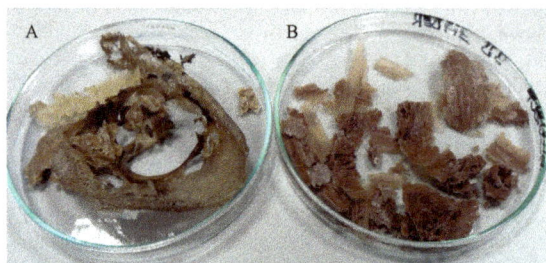

Fig. 1 Dichloromethane extractives content in commercial (A) and studied coil (B).

The commercial and studied coil have a higher extractives fraction than cork due to the presence of PLA. In the case of cork, extractives only include n-alkanes, n-alkanols, waxes, triterpenes, fatty acids, glycerids, sterols, phenols and polyphenols. They are classified into two groups: aliphatics or commonly named cork waxes that are solubilised with low-polarity solvent (e.g. hexane, dichloromethane, chloroform) and phenolics extracted by polar solvents (e.g. ethanol and water) [4; 6].

Nevertheless, in the chemical composition it has been observed that the suberin content of pellets is 25,4%, higher than the commercial (3.6%) and the studied coil (5.8%), respectively. Although lower than cork (30-60%), it is consistent with the percentage of cork in the chemical mix.

The suberin content of the commercial and the studied coil were slightly smaller than expected. Although it was thought that previously extractions were not enough to remove all the PLA, further investigations need to be done. Another extraction in low-polarity solvent might have been necessary to facilitate the extraction of suberin. The lignin and holocellulose content of the pellets (51.5%) was higher than the studied (16.8%) and commercial coil (3.5%), respectively.

According to these results, the pellets chemical composition is similar to cork. Obtaining the pellets was the first step to make the coil, so it could explain that its chemical composition was similar to the cork ones. Following our interest in the development of the cork coil, the commercial coil and the studied cork coil were compared. The amount of cork is an important parameter to take into account in this type of products. According to this, the percentage of suberin is a value related to the cork content: the higher percentage of suberin, the higher the content of cork. In spite of the content of suberin in the studied cork coil being less than expected, the commercial coil had an even lower percentage. Knowing this, the chemical composition suggests that the commercial coil has also a lower content of cork.

Cork Science and its Applications (CSA2017)   Materials Research Forum LLC
Materials Research Proceedings **3** (2017) 11-18   doi: http://dx.doi.org/10.21741/9781945291418-2

The comparison of the FTIR spectra of the PLA coil, the studied cork coil, the commercial cork coil and the pellets did not show obvious differences between them. Although cork spectra seemed divergence between pellets and cork coils, some characteristic bands appeared (Fig 2).

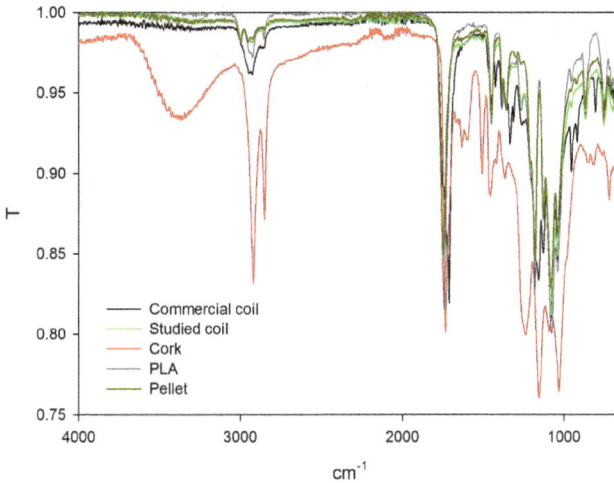

Fig. 2 FTIR spectra of cork, PLA, studied cork coil and commercial cork coil.

In the case of a cork sample, the band $3425cm^{-1}$ indicates the presence of holocellulose and two major peaks of approximately 2919 cm^{-1} and 2854 cm^{-1} correspond to symmetric and asymmetric vibrations respective link characteristic CH_3 aliphatic suberin. Other bands also characteristic of cork are at $1607cm^{-1}$ and $1513cm^{-1}$ corresponding to C = C bond of suberin and lignin and $1162cm^{-1}$ corresponding C-O-C the bond and $1263cm^{-1}$ corresponding C = O the bond of suberin, respectively (Fig. 2). [6-9].

According to figure 2, it would appear that FTIR methodology would allow observing lignin and suberin corresponding peaks ($1162cm^{-1}$ and $1263cm^{-1}$) in commercial and studied coil but these peaks are also characteristics of PLA. Nevertheless, FTIR methodology could not discriminate properly peaks corresponding to cork and PLA. This fact could be explained because the major compound of them is PLA and their principal components may mask the characteristic components of cork.

In order to go in depth, SEM analysis of the PLA coil, the studied cork coil and the commercial cork coil were performed (Fig. 3).

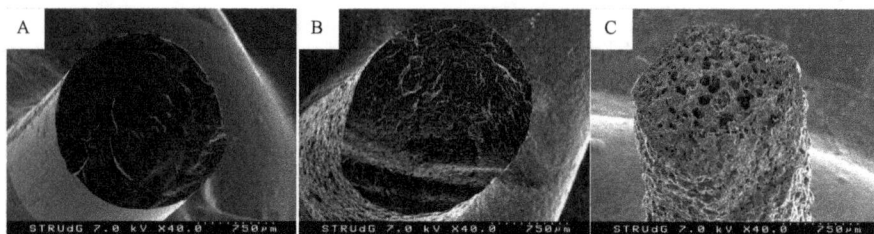

Fig. 3 SEM micrographs of PLA coil (A), cork commercial coil (B) and cork studied coil (C).

According to the micrographs shown in Fig. 3, the morphology of the filaments seems apparently different between them even though PLA (Fig 3 A) is the major component of both the commercial and the studied coil (Fig 3 B and C). The Commercial and the studied cork coil show a more rugged structure compared to PLA. This is probably caused by the presence of vegetal material in its composition.

In the case of commercial and studied coil, it was expected that the typical cellular morphology of cork (tiny hollow hexagonal prismatic cells stacked) [8] were generally preserved. Although, according to figure 3, the typical structure of cork was not observed in either. It could be explained by two hypotheses. On the one hand, most of the cells may be broken during the grinding phase. It should be noted that for the manufacture of the filaments cork powder has been used and therefore cork material had undergone an aggressive mechanical process. On the other hand, during the grinding phase most of the final cork granules had a diameter size of less than the diameter of the cork cell. It should be taken into account that cork powder comprises all cork particles with less than 250 μm. Each cork cell is surrounded by a cell wall and is hollow inside. It is estimated that each cm^3 of cork has 15 to 40 million cells, each measuring approximately 40μm on average [9]. Therefore, some cork particles used in the coils could be smaller than the cork cell.

In addition, differences have been observed between the commercial and the studied coils (Fig 3 B and C). The studied coil presented a rougher morphology than the commercial one probably caused by the different chemical composition and the higher content of cork components (lignin and suberin).

The chemical composition results revealed that the studied coil has higher proportion of cork components than the commercial coil. This fact would be related to the content of cork. An increase of SEM magnifications, allowed to see *some cork particles* dispersed in the PLA matrix in the studied coil (Fig 5B) which were not evident in the commercial coil (Fig 5C).

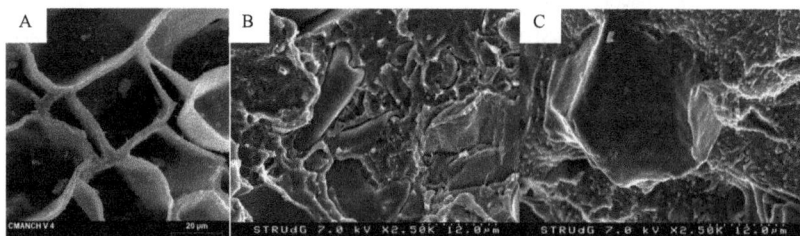

Fig. 5 SEM micrographs of cork (A), cork studied (B) and cork commercial (C).

Print characteristics or features

Both the cork studied coil and the cork commercial coil have about the same thermal durability as PLA and it is actually quite easy to print with both. In the commercial coil, the manufacturer's recommended printing temperature range between 175°C to 250°C [10]. However, we used 165°C as printing temperature and 40°C on a heated print bed. As an interesting side effect, by playing with the temperature settings it could be avoid that a higher temperature will create a darker color and burnt smelt. Another aspect to consider is the printing speed as it is recommended not to use very high speeds with such filaments "experimental material status". So we used 130m/s intend of 160m/s (height print speed recommend in PLA). In this case we have also obtained exceptional results (Figure 5) preserving the high level of detail of the original model. Notwithstanding that the commercial coil presented more flexibility than the studied coil, the material has a greater flexibility than the PLA in both impression for both cork coils.

Other physico-mechanical characterization of the coil will be done according to international code.

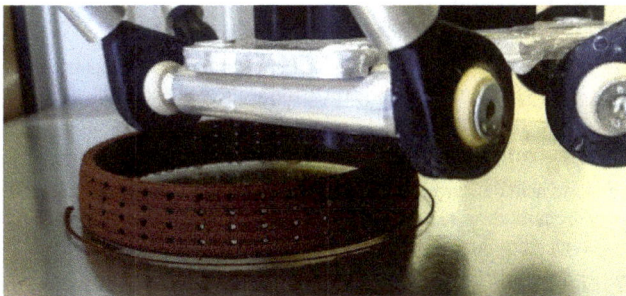

Fig. 5 Model of printing a cork studied coil.

Conclusions

Novel duel coil base in cork was successfully obtained. In general, the cork studied coil showed more content of cork than the cork commercial coil. The presence of cork material in a 3D filament can be evaluated using a chemical composition assays and SEM. The percentage of suberin is related to the cork content and allows us to compare the differences of these cork products. FTIR is not a good methodology to assay PLA based products.

This work represents, therefore, very interesting material for the up-grading of an important industrial residue and simultaneous for the development of a new class of sustainable material with potential application in areas such as in the automotive and decoration industries. Work is in progress to further evaluate other features of this novel sustainable material and future biocomposites should be explored.

Acknowledgments

The authors thank the Department of Territory and Sustainability of the Catalan Goverment and the Waste Agency of Catalonia for the grant of the project in the framework of the Circular Economy TES/1275/2016. The authors especially thank 3D Spider Print (http://3dspiderprint.com/index.php) for the production and development of the studied cork filament.

References

[1] Allied Market Research 3D Printing Market by Tecnology (Stereolithography, Selective laser sintering, Electron beam melting, Fuesed deposition modeling, Laminated object manufacturing) and Material (Polymers, Metals & Alloys, Ceramics)- Global Opportunity Analysis and Industry Forecast, 2014-2020. (2015) Consulted: http://www.alliedmarketresearch.com/life-sciences-market-report

[2] EIP-AGRI Inspirational ideas. Cork fir 3D printing. Newsletter on Agriculture & Innovation. (2017) Edition 44

[3] Rives J. et al. Environmental analysis of cork granulate production in Catalonia-Northern Spain. Resources, conservation and Recycling. Elservier: (2012) 58 132-142.

[4] Jové P. et al. Study of the variability in chemical composition of bark layers of Quercus suber L. from different production areas. BioResources (2011) 6 (2): 1806-1815.

[5] Pereira H. Chemical composition and variability of cork from Quercus suber L. IAWA Bulletin n.s., (1988) 8 (3).

[6] Pereira H. Cork: Biology, Production and Uses. Elevier. (2007)

[7] Pintor et al. Use of cork powder and granules for adsorption of polluants: a review. Water Res., (2012) 46; 3152-3166. https://doi.org/10.1016/j.watres.2012.03.048

[8] Vilela C. et al. Novel sustainable composites prepared form cork residues and biopolymers. Biomass&Bioenergy. (2013) 55: 148-155. https://doi.org/10.1016/j.biombioe.2013.01.029

[9] Jové P. Caraterització del suro i subproducte de la indústria surera. Valoració d'aquests com a biosorbents d'hidrocarburs aromàtics policíclics en aigües d'escorrentia. (PhD Thesis). (2015) Universitat de Girona, Girona.

[10] Form Future Easy Cork. The building block of your creations. (2016) Consulted: http://www.formfutura.com/285mm-filaments/easycork/

[11] Wise, L.E.et al. Chlorite holocellulose, Its fraction and bearing on summative wood analysis and on studies on hemicelluloses, Pap. Trade J. (1946) 122: 34-43.

Cork Science and its Applications (CSA2017) Materials Research Forum LLC
Materials Research Proceedings 3 (2017) 19-26 doi: http://dx.doi.org/10.21741/9781945291418-3

Designing with Cork – Bringing the Industry into the Design Studio Classroom

T. Franqueira[1a], J. Sampaio[1b], E. Oliveira[1c], C. Pereira[1d], A. Kumagai[1e]

[1] University of Aveiro, Department of Communication and Art, Campus Santiago, 3810-193 Aveiro, Portugal

[a]teresa.franqueira@ua.pt, [b]joao.sampaio@ua.pt, [c]emanueloliveira@ua.pt, [d]catiapereira2@ua.pt, [e]kumagai@ua.pt

Keywords: University-Industry Collaboration, Cork, Product Design

Abstract. This paper presents a collaborative project between the Product Design Course of the University of Aveiro and Amorim Cork Composites. This project was carried out during 2 months, and aimed to generate concepts and products for office and domestic environments based on the raw material of the company – cork. Using the Double Diamond methodology[1], we implemented the four phases (discover, define, develop, deliver) throughout the project. During the discovery phase, the students had contact with the company at two different moments: in the first one, the company gave a presentation in the classroom and on the second meeting, the students visited Amorim's production unit and showroom. The students were introduced to the company's products and production processes and technologies allowing the creation of awareness of the specificities of the material that would be used in their projects. Afterwards, in the define phase, students gathered insights and information about the technical features of the material and the possibilities to apply cork in different contexts. The students started by combining this information with potential user scenarios, developed concepts and finally refined their brief. The brief validation occurred during the third contact moment with the company, which was a presentation session in the classroom. The development phase began with exploration and ideation, leading to the identification of possible solutions and converting concepts into tangible ideas. As the projects evolved, the solutions were tested as mock-ups that allowed for the first experiments and evaluation. Adjustments were made by the students, and their focus shifted to the final phase of delivery, and the subsequent need to prototype with the material provided by the company. The fourth and final contact moment with the company enabled the students to present their final prototypes and project with detailed information. The relationship between the company and the university allowed students to develop concepts and products for the real market context, taking into consideration specific technological and productive constrains and to create solutions for a specified client with defined needs. Furthermore, this project brought to discussion the relevance of including specific challenges within the pedagogical component for the students training, allowing academia to evolve to a more open and collaborative relation with local companies. Another aspect addressed was the development of solutions derived by the material's properties (taking advantages of the unique features of cork), made the results emerged from a research push context instead of a market pull one.

Introduction

The region of Aveiro is characterized by the presence of a large number of industries which cover many areas of expertise, serving as a main cluster for several production sectors, although in the present it doesn't have the same impression as it had 20 year ago.

---

[1] http://www.designcouncil.org.uk/news-opinion/design-process-what-double-diamond

Cork Science and its Applications (CSA2017)   Materials Research Forum LLC
Materials Research Proceedings **3** (2017) 19-26   doi: http://dx.doi.org/10.21741/9781945291418-3

The University of Aveiro is located at the core of this environment and assumes as one of its main missions for the region to generate networks that bring together in collaborative projects academia with the social and economic fabrics of the region. These partnerships and collaborations are part of a solid, promising and prestigious platform that intends to generate social valorization through the transfer of scientific knowledge. The objective is to look for new formats that enable the university to assist the surrounding social and economic fabrics through its scientific and technological competences.

The *Master's Degree in Engineering and Product Design* (MSc) is a platform with enormous potential to explore these company-university interactions, given its practical nature, and its embedded notion of real production constraints. This MSc is an educational program that is focused on provoking innovation through design, and intends to shorten the industry/university gap and claims itself as a reliable factor for the rejuvenation of Portuguese companies.

The textiles, footwear, and cork industries are now seen as the main national engines of innovation[2], in these sectors, Portuguese companies when not leading, they match the main players, being among the best that is done worldwide. Thus, being the design studio course the backbone of this educational program (MSc), a big proximity between companies that are world leaders of innovation and young creative minds presents itself as win-win situation for all stakeholders.

Cork as a raw material and as an identity

Before the designing process start, designers should have a deep understanding of the full specifications and constraints that give a social, economic, technological and cultural context for the intended solution.

Cork is the outer layer of the cork oak (Quercus suber), making it a 100% natural vegetable tissue, it is formed by a honycomb of microscopic cells filled with an air-like gas and coated mostly by suberin (highly hydrophobic and a somewhat 'rubbery' organic material) and lignin (organic polymers). This natural, versatile and sustainable raw material is so perfect that until now now synthetic process has been able to mimic it's unique properties: Lightness and Floatability – more than 50% of its volume is air; Elasticity and Compressibility – with a strong elastic memory, it's the only solid that when tightened on one side, does not increase its volume on the other; Impermeable to gases and liquids; Excellent thermal and acoustic insulation capacity; Fire retardant; high resistance to friction; hypoalergenic; soft touch. (referencias)

Have been used for centuries with many different applications, cork is core to the genesis of the endogenous Portuguese forests (ocuping more than 20% off national forests), and it constitute a significant element of its cultural heritage — a essential aspect to be considered by the students: all projects should explore esthetic, mechanical and symbolic aspects of cork. Although in the last decade the cork industry has grown exponentially and has been asserting itself in many new markets (mainly through cork stoppers), there is a general intent to explore new, more innovative opportunities and with higher added value, where design has a key role.

Being the Master in Engineering and Product Design lectured in the Aveiro Region, that is very close to the major Portuguese cork companies it becomes more and more relevant to generate synergies that generate innovative products with this unique raw material. Fostering collaborations between the industries that have the market end technical know-how, the university that can offer fundamental investigation, the cultural and social tissues that guarantee cultural continuity and the MSc as place for materialization is the main objective.

---

[2] https://www.dinheirovivo.pt/economia/sectores-tradicionais-de-obsoletos-a-casos-de-sucesso/

Based on this framework, the teacher responsible for the Product Design Studio Course established contact with Amorim Cork Composites to develop a collaboration agreement for product development that considers cork as the main raw material. To develop design-led projects that explore cork's aesthetic potential, its technical features as well as its symbolic and emotional characteristics for the retail sector was the out frame of the collaboration.

The opportunity and development of this collaboration between the university and the cork industry is motivated by the idea that innovation can outcome from the exploration of a traditional material, but it is also driven by other factors: the cultural, economic and social impact at regional and national level of cork industry (with a trade balance of 757 M €)The stimulation of proximity between the University of Aveiro and the local industrial clusters; The specific characteristics of the syllabus of the MSc in Engineering and Product Design of the University of Aveiro; And lastly the pragmatic and hands-on approach that is given to the double diamond design process, by keeping in mind real social and industrial contexts advocated in Product Design studio of the above-mentioned master course .

Project definition
In order to guarantee successful results, the project brief would have to take into consideration the companies interest and cross these with the courses pedagogical and programmatic established outlines. Preliminary meetings took place where both parties shared motivations and interests that led to the definition of a global theme and specific areas to be addressed, at the same time keeping in mind that the project had to be inspiring and motivating for the students. Bringing the company in when defining the project brief intended to manage their expectations for the project, and the opportunity for the lectures to clarified the specific programmatic needs that had to be assured in this specific learning moment. Despite the nature of this project being practical, the moments of interactions between students and the company were to consider educational aspects prior to the company's interests. After creating a shared theoretical and conceptual framework, it was easier to define the theme and nature of the challenge to be presented to the students.

Students should explore new opportunities for cork by exploring and highlighting its natural and mechanical characteristics and its emotional cultural and sensorial specificities. Students would be asked to choose one out of the two defined contexts:

a.   Design a set of cork products for the domestic environment (1st theme).

b.   Design a set of cork products for work / office environment (2nd theme).

For both contexts, students were encouraged to propose and explore the unique characteristics of cork, with the intent that this would lead them to original approaches to every-day artifacts. The outcome should be artifacts that would be perceived in the market as innovative and with added values though the material exploration, as Fallan (2012:42) explains — "The material itself conveys messages, metaphorical and otherwise, about the objects and there place in a culture". In the development process, they should relate and constraint the main material-cork (and complementary materials), to the existing manufacturing processes made available by the company and functionalities required for the defined contexts. The project should result in a set of artifacts that establish a new aesthetic language for the use of this material, a new functional approach, creating or adapting to new behaviors arising from contemporary lifestyles of the given contexts.

It was agreed with the company that the students would have the freedom to choose which of the proposed context they would want to develop. These proposals would be developed in groups of two students that should complement each other (having one a more technical approach to problem solving and the other with a more conceptual one). As a result of this organization strategy with the final group layout was of 3 groups that selected the 1st theme (domestic environment) and 3 groups for 2nd theme (office environment).

The design methodology presented and recommended for students to approach their design problem was the "Double Diamond" in order to enhance several moments that support divergent thinking and convergent thinking that foster creative and innovative solutions.

Discover

The project was presented to students in two presentations. Being the first exposed by the teachers, which launch the challenges and enhance a debate and reflection about how we live in domestic and work office contemporary environments, new societal behaviors and to look at cork as what it could be more than what they know it is. This material exploration and behavioral-led design research allowed students to gather information that would sustain their future scenarios.

A second presentation was given by Amorim Cork Composites (Figure 1), who relayed on technical constraints: the company's structure, its market positioning, the unique mechanical properties of cork, technical limitations, the diversity of products and applications exploited by Amorim RD team and among other aspects, their main clients. This presentation also reflected the company's concerns on sustainability and circular economy, creating awareness for the importance to look at cork as a natural recourse that should be valued and used efficiently.

Fig. 1 Presentation by Amorim Company.

After this first introductory moment, the first phase *(discovery)* was initiated. This phase is divergent and exploratory, searching to establish new user needs. In this moment, students are encouraged to look at the world in a different way, questioning everything, gathering all types of information in a non-filtered manner, looking at things as if it was the first time, and to gather insights from their perspective. With this in mind, a visit to the company (Figure 2) was organized, with the intention of giving students the opportunity to see and question production process and not blindly follow the recommendations given by the company. In this complementary activity, the students had the chance to contact directly with a wide diversity of different types of cork, cork composites, methods and form-giving technologies of the various materials (Figure 3).

Cork Science and its Applications (CSA2017)   Materials Research Forum LLC
Materials Research Proceedings 3 (2017) 19-26   doi: http://dx.doi.org/10.21741/9781945291418-3

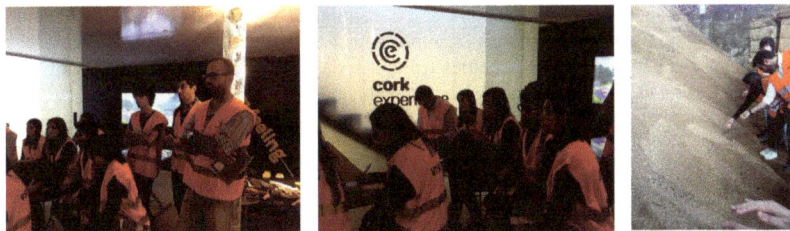

Fig. 2 Visit to Amorim Company.

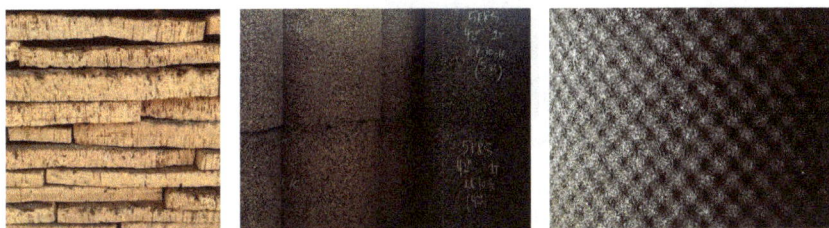

Fig. 3 Cork material by Amorim.

Define

In the second phase *(define)*, in which designers try to make sense of all data collected and the possibilities identified, the goal was to develop a clear creative brief that frames the fundamental design challenge, at this point students were mentored to identify opportunities and to generate their unique interpretation of their findings for further consideration. To validate their intended approaches, students explored mainly corks technical features searching for new possibilities where to apply cork in new contexts that have been undervalued until this moment. Subsequently, they started to combine this information with the behavioral aspects and potential new niche user groups, creating mood boards as a way to better define and communicate their ideas. Although in this phase the intent is to converge into a defined project definition, students where guided to keep an extended range of ideas that would be better defined in a contact moment with the company. In this phase they were asked to describe their concepts by 3 identifying aspects: context, user, and innovation and relevance for the application of cork.

The brief validation occurred during the third contact moment with the company (Figure 4), which was characterized by an informal presentation/talk in the classroom to the Amorim's marketing and manufacturing staff. This informal talk between each group was mediated by one of the three lectures in the classroom.

Based on the first concepts developed, this contact had several goals and outputs:

a.   The presentation of the concepts developed;

b.   Analysis of conceptual potential, adequacy to user and markets;

c.   Analysis of productive viability by the company's retail team;

d.   Clarification of doubts and technical specifications, normative and productive issues;

e.   Stabilization and definition of a brief.

Cork Science and its Applications (CSA2017)   Materials Research Forum LLC
Materials Research Proceedings 3 (2017) 19-26   doi: http://dx.doi.org/10.21741/9781945291418-3

This contact with specialized interlocutors of the manufacturing and marketing team from the company, allowed the students to frame their proposal within the Amorim Cork Composites philosophy and by integrating the companies technical know-how avoided the persuit of ideas that were not meating the fabrication and systemic reality.

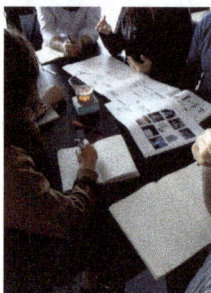

Fig. 4 Third contact moment with the company.

This momentum proved to be extremely important, making the companies personnel think out of the box, and the students to validate concepts with real active players of the professional market. Subsequently, the students developed a refined proposal that reformed and adapted the project's specifications based on the information and feedbacks that emerged from the contact with the company staff members.

Develop
With the previous phases finished, and the consequent brief definition that resulted from it approved, the students began to development a conceptual framework for their proposals. The third phase *(develop)*, marks the second divergent moment of this process, in this stage ideation is catalyzed through intense sketching, brainstorming, reverse brainstorming exercises and lo-fi mockups. This process of trial and error helped students to improve and refine their ideas. In this phase it's very important for students to keep in mind all the data taken into consideration in the previous phases, managing students enthusiasm by constantly asking for validation of their concepts through their previous findings is vital for a satisfactory outcome. Looking for a formal identity that has the designers unique point-of-view and that is coincident with market demands, led to benchmarking market trends analysis and user's type studies. These subjective options were complemented with technological constraints and the materials' identity. Although still in an embryonic and exploratory phase, all projects started to blossom into valid proposal for new innovative products that answered to the defined and validated brief. For a better understanding and exploration of concepts, lecturers suggested the students to develop low-fi mock-ups in order to rapidly test many different solutions. This task is related not only to the need for shape exploration, but also to understand potential fabrication difficulties and limitation of the raw-material. In addition, visualizing design options worked as a step stone to enhance the next convergence phase by consolidating all data and ideas as a single proposal.

As a conclusion to this phase, a final low-fi physical representation of each proposal was developed and presented as a way to validate the logic behind the proposal and the productive feasibility of the artifacts. Parallel to this task of refinements and final technical adjustments of the proposals, the students were informed of the importance of developing the branding and communication strategy in order to obtain better results. As an approach to enhance the semantic

Cork Science and its Applications (CSA2017)   Materials Research Forum LLC
Materials Research Proceedings **3** (2017) 19-26   doi: http://dx.doi.org/10.21741/9781945291418-3

profile of the project, a series of tasks were requested from the students. In addition to a functional prototype, they were asked to develop virtual models in order to test finishing and simulate the placement of the artifacts in different contexts. Students started to developed its graphic: name and logo; communication strategy; packaging, and other elements that could give advantages in all phases of the product chain (production, distribution, sale, use and recycling). All these contents were to be included in a booklet that present and resumed the overall project:

a. concept;

b. process;

c. branding and communication strategies;

d. images that simulate the artifacts (virtual or real) in usage scenarios and contexts;

e. constructive technical data and drawings.

All the elements referred above were included in the final presentation in the format of an elevators pitch as well as multimedia presentation to the teachers and colleagues. After a formal 6-minute presentation of each group an open analysis and debate was carried out, where the teachers and all students could give feedback on the proposal with the intent to improve the projects before the final step.

Deliver (Final presentation to the company)
The final stage of the methodology (*deliver*), is where the resulting project (a product, service or product service system) is finalized, prepared for industrialization and ready to be market tested. First, a working prototype was delivered to the company to be market tested in small test groups, where real market feedback is gathered, and taken into consideration for final adjustments. Although the final *deliver* to the market wasn't possible, the presentation to the company was seen as a possible preliminary first step to this stage.

After the course's final presentation, the opportunity was given to the students to improve their project and final prototype, and this was done with the intent of guaranteeing a more professional outcome to be presented to the company.

The students' final projects presentation to the company took place after the end of the semester at the university in the format of an open exhibition. With this action, the aim was to create a moment of debate and analysis between the students along with Amorim Cork Composites marketing and production staff based on the perceived market potential of the proposal, their quality and productive feasibility. For that an exhibition stand was set up as a project presentation strategy (Figure 5).

For this final pitch, the students also used a multimedia presentation, a booklet with conceptual and technical details of the work as well as the prototypes of the products. Each group has approximately 6 minutes for presentation, followed by 6 to 10 minutes of discussion with the representatives of the company.

Since the market implementation of the products was not possible to be developed in the classroom context, this ongoing work is being driven by the company, and the R&D team worked with the students on refining the proposals in a logic of generate greater value for the company. Once all the projects were submitted to the company, their feedback was that in most projects enough potential was identified to justify further investment. The option that this final design phase should be continued within the Amorim Cork Composites company was taken because it made more sense since they already have established the appropriate distribution channels.

Cork Science and its Applications (CSA2017)   Materials Research Forum LLC
Materials Research Proceedings **3** (2017) 19-26   doi: http://dx.doi.org/10.21741/9781945291418-3

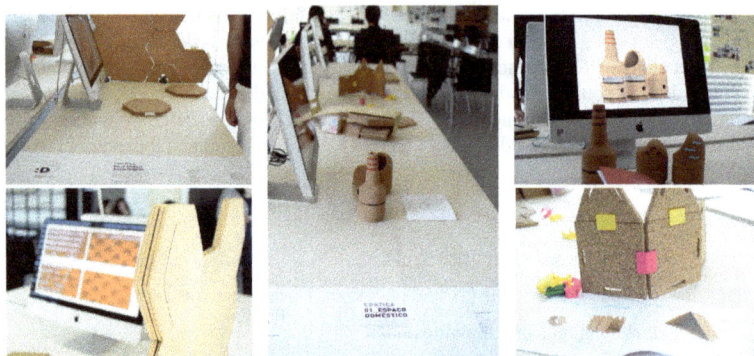

Fig. 5 Final presentation and exhibition

Conclusion and discussion

Cork presents itself as a raw material with enormous potential, not only for its unique mechanical characteristics, but also presents great symbolic value that can be explored as a part of the Portuguese cultural heritage.

The opportunity to define a design challenge in collaboration with partners like Amorim Cork Composites that deal with real needs and not hypothetical ones, proved to be a good opportunity to create awareness to the students on issues related to the need of rethinking Portuguese culture, as well as it promotes a proximity between academia and the industry.

The use of the Double-Diamond methodology, helped to plan and coordinate the various contact moments between the students and the company, and proved to be an effective approach in tutoring projects. In addition it allowed students to approach design problems in a methodical way, giving them a clear notion of where they are in the design process, with very defined goals without losing creativity and a sense of authorship although collaborating with other people.

Of the six projects presented to the company's team, they identified 3 to 4 project with market and productive potential and that the company will take to the final implementation stage. This scenario indicates that university-industry collaborations can be very fruitful for both parties.

References

[1] Double Diamond Design Process Model (Design Council, 2005) In:
http://Www.Designcouncil.Org.Uk/ Designprocess [Acedido Em: 27 May 2015]
Fallan, K. (2012). Design History: Understanding Theory and Method. New York Berg.

[2] Silva, S. P., Sabino, M. A., Fernandes, E. M., Correlo, V. M., Boesel, L. F. & Reis, R. L. (2013). Cork: Properties, Capabilities and Applications. International Materials Reviews, 50(6), 345-365.

[3] 1001 Negócios Da Cortiça (Pereira, J And Cardoso, M. 2014) In
(Http://Expresso.Sapo.Pt/Economia/Exame/1001-Negocios-Da-Cortica=F873107) [Retrieved on September 18th, 2016]

Cork Science and its Applications (CSA2017)
Materials Research Proceedings 3 (2017) 27-31

Materials Research Forum LLC
doi: http://dx.doi.org/10.21741/9781945291418-4

Cork in the Field of Renewable Energies

Luís Gil

¹Direção Geral de Energia e Geologia, Divisão de Estudos Investigação e Renováveis, Av. 5 de outubro, nº208, 1069-203 Lisboa, Portugal

Keywords: Cork, Renewable Energy, Cork-derived Materials, Biomass, Solar Energy, Wind energy, Bioenergy

Abstract. This work describes the multiple ways in which cork is now related with renewable energy, and other future possibilities. Cork wastes can be used as biomass sources for energy production and cork-derived materials can have several applications in different renewable energy production or utilization systems. Several types of renewable energies can be used in different steps of cork processing. Some clues are also pointed out for possible future applications of cork in renewable energy systems as well as new uses of renewable energy technologies in cork industrial processing.

Introduction

Cork is obtained from the bark of the cork oak (*Quercus suber* L.) and is one of the most important non-timber forest products. It is mostly produced in the western Mediterranean countries, Portugal being the leader producer (34% of the world's area of cork forests and 49,6% of cork total production). World cork production reached about 201 000 tonnes in 2014 and the cork oak forests cover a worldwide total area of just over 2.1 million hectares [1].

Cork is world known as the material of excellence for the sealing of alcoholic beverage bottles. But this natural cellular material has also many other interesting applications. Portuguese cork exports are led by cork stoppers, accounting for about 70% of the total value and cork building materials account for about 26% [1].

The cork structure may be described as an array of closed prismatic hexagonal (on average) cells stacked base-to-base, forming a honeycomb-type structure. The cork cell walls show some undulations of varying intensity. Cork has also a layered structure (annual growth rings), with different cork cells depending on the period of formation (different height and different wall thickness). The solid cell wall content is only about 10% and the rest is retained air. The cell walls are suberified. All this contribute to the characteristic properties of the cork material which allows it a wide world of applications [2].

Considering now the relationship between cork and renewable energies, one can say that several renewable energy production systems and components are using or can use cork and cork-derived materials. Besides this some renewable energy technologies are or can be used in several operations of the cork industrial processing.

Bioenergy from several cork processing wastes

Cork powder, the major cork processing waste, is used as a biomass fuel in boilers for heating water in the cork boiling process, and for steam production in the insulation corkboard agglomeration process and also in several heating steps in cork stoppers and composite cork manufacturing [3]. This waste has a good calorific value [2] of about 4000 to 7000 kcal/kg, depending on the type of powder and is usually burned in conventional burners.

Cork Science and its Applications (CSA2017)   Materials Research Forum LLC
Materials Research Proceedings 3 (2017) 27-31   doi: http://dx.doi.org/10.21741/9781945291418-4

Another industrial waste is the cork boiling wastewater. This liquid waste has to be treated and this treatment can be carried out by anaerobic digestion in order to produce biogas which can be used in other cork processing operations [4], namely cork boiling for cork stoppers production.

Fig. 1 Cork boiling wastewater.

This biogas/biomethane can also be used, e.g. in cork stoppers drying, cork stoppers or other cork items fire printing or to generate the processing heat needed in the cork composites agglomeration process.

Cork in solar energy applications
A link between cork and solar energy can be the heating of water in order to produce boiling water (~100°C) for the boiling of reproduction cork, the first industrial processing step in the cork stoppers manufacturing. Steam production (>300°C) for the agglomeration of cork granules in the insulation corkboard production is also a possibility for the use of solar energy in the cork sector. In the first case medium-high temperature collectors can be used, and in the second case it is foreseen the use of concentrated solar technology.

Solar energy can also be used in solar kiln drying of cork raw material, namely cork material with excess of moisture.

Cork can be used as a matrix for the production of biomimetic cork-based ceria ecoceramics in order to obtain ceramic foams having the cork cellular structure, for hydrogen generation using concentrated solar energy [5]. Hexaferrites can be used in order to obtain magnetic ceramic foams by pyrolysis, using cork as matrix and template, resulting in a very light and porous solid with the cork microstructure. The **directly produced hydrogen (water splitting),** as an **energy vector,** can increase the penetration of **renewable** and intermittent sources in the **energy** supply.

Pipe insulation with flexible extruded composition cork is used e.g. in thermal solar energy, due to weather and ageing resistance of this product when in outdoor use [6]. Solar thermal collectors produce hot water but this water has to be transported from the heat storage zone to the tap. At least part of the water piping is in the exterior, where the weather factors (UV radiation, rain, frost) accelerate the pipe thermal insulator degradation. This cork agglomerate pipe insulation is flexible, easy to handle and to apply, and has been demonstrated as very resistant to the weather action, decreasing maintenance costs. Of course this is also true for the piping of any kind of hot fluids, of any renewable energy systems, with a limit of about 200°C.

Cork Science and its Applications (CSA2017)   Materials Research Forum LLC
Materials Research Proceedings **3** (2017) 27-31   doi: http://dx.doi.org/10.21741/9781945291418-4

Fig. 2 Extruded cork composite for pipe insulation.

The so-called "projected cork" is another product containing cork that can be used in solar applications [7]. Particles of cork are mixed with a polymer/binding material, usually in situ, and this mixture is spray-gun projected directly into the surface to cover, in an easy way. The temperature range of work is from -165°C to +165°C. The mixture sticks well to any kind of surface. It is a material with great elasticity, flexibility and mechanical resistance. It can be used to repair some defective insulating material previously existing. Projected cork can also be used for insulating the reservoirs and/or piping of solar thermal collectors or other renewable energy systems needing protection in the fields of impermeabilization and thermal and acoustic insulation.

Cork-derived material is also foreseen for the production of solar thermal collectors, constituting the collector body with thermal insulation capacities. Using adequate cork composites, it is possible (some exploratory experiments were carried out) to form, in one step, the solar thermal collector body, incorporating some stiff and mechanically resistant support systems to allow the assembling of the collector. The foreseen technology is based on high diameter metal tubes which work as water reservoirs and heaters.

Fig. 3 Solar thermal collector body made of cork composite.

Cork Science and its Applications (CSA2017)   Materials Research Forum LLC
Materials Research Proceedings 3 (2017) 27-31   doi: http://dx.doi.org/10.21741/9781945291418-4

Cork applications in wind energy

As cork is known as a very adequate friction material, there is the possibility to use cork brake pads, e.g. to apply in small wind turbines brake systems. Small wind turbines usually require a simple rotational speed control system and brakes are needed not only to ensure a safe operation, but also to keep it within its working range. There are many brakes in the market, but the vast majority is not suitable for the variable torque/speed characteristic of this equipment. Problems in their long term utilization have been identified, mainly for the low rotation speed regimes. In order to optimize the brakes behavior, the use of cork - as a natural and sustainable friction material - in the composition of their brake pads is foreseen due to its resilience and friction behavior.

Cork anti-vibration insulation materials, based mainly on high density insulation corkboard or composition cork (e.g. with rubber), can be used in vibration insulation in wind towers and other energy systems [8]. Wind towers are under mechanical stress due to wind velocity changes and, in order to attenuate wagging, some vibration absorbing materials can be used, namely cork-based materials (pure or composite).

It is worthy to mention also that some experiments were carried out using cork material as a component in the rotor blades of the wind turbines in order to improve the acoustic and vibrational performance [9]. The study on rotor blades material in order to find cheaper materials, without significant performance problems, drove the research to sandwich structures with cork core composites as a rotor blade material [10].

Cork in full and hybrid electric vehicles

The combustion engines working with biogas/biomethane, used in conventional vehicles or hybrid vehicles use cork based gaskets and joints, namely from corkrubber or rubbercork (depending on the v/v ratio cork:rubber) composite material in several engine and transmission parts.

New cars based on fuel cells, batteries or other electric engine based type, tend to have a flat-bottom body, which is a good characteristic for cork floor applications in this field decreasing weight and carbon footprint [9].

Other cork uses in the renewable energy field

Cork based gaskets and joints have also several applications in several component parts of renewable energy systems, whenever a sealing or expansion damping material is needed.

As cork is also very resistant to salt water the utilization of this material in off-shore technologies (e.g. wind, wave energy) can be a future field of research.

Conclusions

Cork is a very versatile material having very different applications even in the field of renewable energies. Furthermore, cork can also be related with renewable energies in its processing steps. New cork applications in this field may be foreseen in the future.

References

[1] Cork Yearbook 2015, Portuguese Cork Association, Ed APCOR, Santa Maria de Lamas, 2015.

[2] L. Gil, Cortiça: Produção, Tecnologia e Aplicação, Ed INETI, Lisboa, 1998.

[3] L. Gil, Cork powder waste; an overview, Biomass and Bioen., 13 (1/29), 59-61, 1997.

[4] I. Marques, L. Gil, Energetic potential of cork processing wastewaters, Int. Cong. Water, Waste and Energy Manag., p. 43, Salamanca, 2012.

Cork Science and its Applications (CSA2017)   Materials Research Forum LLC
Materials Research Proceedings **3** (2017) 27-31   doi: http://dx.doi.org/10.21741/9781945291418-4

[5] R. Pullar, L. Gil, F. Oliveira, Biomimetic cork-based ecoceramics for hydrogen generation using concentraded solar energy, MatCel'2015 – 1ª Conf. Materiais Celulares, Aveiro, 2015.

[6] D. Esteves, Desenvolvimento, Caracterização e avaliação do comportamento termo-mecânico de um novo material constituído principalmente por cortiça, MSc thesis, Instituto Superior Técnico, 2010.

[7] information on http://www.subertres.com/

[8] H. Policarpo, A. Diogo, M. Neves, N. Maia, A note on the estimation of cork composite elasto-dynamic properties and their frequence dependence, Proc. Int. Conf. Struct. Engineering Dynamics, Sesimbra, 2103.

[9] L. Gil, New Cork-based Materials and applications, Materials, 8 (2), 625-637, 2015. https://doi.org/10.3390/ma8020625

[10] S. Kim, Study on Mechanical Properties of Cork Composites in a Sandwich Panel for Wind Turbine Blade Material, MSc, Thesis, Massachusetts Institute of Technology, 2009.

Cork Science and its Applications (CSA2017)   Materials Research Forum LLC
Materials Research Proceedings 3 (2017) 32-39   doi: http://dx.doi.org/10.21741/9781945291418-5

Layer Thickness of Cork Accessible to Extraction

Yaidelin J. Alves Manrique[1], Paula Rodrigues Pinto[2], Manuela V. Oliveira[3] and José M. Loureiro[1*]

[1]Associate Laboratory LSRE-LCM. Faculdade de Engenharia da Universidade do Porto (FEUP), Universidade do Porto. Rua Dr. Roberto Frias, 4200-465 Porto, Portugal

[2]on leave to Forest and Paper Research Institute – RAIZ, Aveiro, Portugal

[3]Instituto Politécnico de Viana do Castelo. Av. do Atlântico, 4900-348 Viana do Castelo, Portugal

*corresponding author: loureiro@fe.up.pt

Keywords: Granulated Cork, Powder Cork, Soxhlet, Extraction

Abstract. The aim of this work is to evaluate the layer thickness of cork accessible to extraction for different solvents. Samples of granulated industrial cork (up to 6 mm) from milling punched cork planks were extracted with solvents of different polarities. The best yield was ca. 96 $g_{extract}/kg_{Cork}$, obtained for the more polar solvent tested, for the smallest particle size; it was assumed that these particles were completely extracted. For particles bigger than 1 mm, the extracted layer is nearly constant for each solvent; for hexane, this extracted layer is around half that of the extracted with the other two solvents, regardless of particle size. The dimensionless thickness, on average, is ca. 0.22, 0.20 and 0.11 for acetone, dichloromethane and hexane, respectively.

Introduction

Cork is a novel material composed by suberin, lignin, polysaccharides and extractives (5-16%) [1-4]. The extractives can be removed by solvent extraction and are mainly constituted by triterpenes as friedelin, betulin, betulinic acid, cerine and β-sitosterol; these compounds were studied and it was found some promising applications as isolated compounds. The authors have been studying the possibility to use cork as a source of these triterpenes, without compromising the future application of granulated cork and cork powder.

It is known that cork has unique properties that give rise to a wide range of applications. However, until now, the main use of cork is still, with no doubt, the production of cork stoppers mainly for the wine industry [5].

The first step in the production of natural cork stoppers consists in the harvesting of cork oak bark, that occurs each nine-twelve year, according with the legislation of each country. After the harvesting process, the cork planks are left out in the open air for a stabilization period and are then sorted by quality and size. Next, the planks to produce stoppers are boiled in purified water free of any microbiological contamination, by ca. 30 min. Afterwards, they are submitted to a quality selection process and kept in controlled ambient conditions. After a stabilization time, the planks are ready to continue the production process and selected planks are punched, giving rise to the natural cork stoppers. In this stage, all material unsuitable to be used in cork stoppers is granulated and separated by particle size and density according to its following applications. It should be highlighted that in the cork industry it is possible to take advantage of all the material, i.e., *"nothing is waste"*; then, the granulated material can be used in the production of technical cork stoppers and as agglomerates, among many other applications.

Cork Science and its Applications (CSA2017)   Materials Research Forum LLC
Materials Research Proceedings 3 (2017) 32-39   doi: http://dx.doi.org/10.21741/9781945291418-5

In order to add a new value to cork, enlarging its markets, the raw material used in this work, as source of the above mentioned triterpenic compounds, was constituted by the granulated and powder generated in this stage.

The aim of this work is to determine the layer thickness of cork accessible to extraction, using different solvents, in order to evaluate the potential to use granulated industrial cork with several particle sizes as a source of triterpenic compounds. This work is important, since it constitutes a necessary step in the evaluation of the maximum profit that can be obtained without modifying the structure of the raw material, given that in the future the raw material extracted should be used in its traditional applications.

Materials and Methods

As previously mentioned, the granulated industrial cork and powder cork used in this work were obtained from milling punched cork planks (after the cooking stage). The samples were separated by visual inspection in two different types: (i) granulates from outer sections of the plank (A7, A9), that are characterized by a darker colour, and (ii) from inner sections of the cork plank (A1-A6); sample A8 corresponds to cork powder. Fig. 1 shows to which section belongs each sample, i.e., inner (A1 to A6) or outer section (A7 and A9) of cork planks.

Fig. 1 Granulated industrial cork from two sections of cork planks: inner and outer.

Table 1 shows the average particle size for each sample, as well as its moisture and density. The moisture of each sample was quantified as follows: a solid sample (ca. 6-7 g) was introduced in a cylindrical container, weighed and placed in an oven at constant temperature (104°C); after 24 h, the sample was re-weighed and placed again in the oven until constant weight. The initial and last weights were used to determine the moisture of the sample. In its turn, to evaluate the cork (solid) density, helium pycnometry was used.

Table 1: Average particle size of industrial granulated cork samples [6]

Sample	Size group [mm]	Average particle size [mm]	Moisture [%]	Cork density [kg•m^{-3}]
A1	0.5 – 1	0.75	5.4 ± 0.1	710 ± 58
A2	0.5 – 2	1.25	6.3 ± 0.1	334 ± 42
A3	0.5 – 1	0.65	5.0 ± 0.1	546 ± 102
A4	1 – 4	2.5	6.7 ± 0.1	326 ± 21
A5	4 – 6	5	4.8 ± 0.1	266 ± 14
A6	2 – 4	3	9.2 ± 0.1	267 ± 27
A8*	< 0.2	< 0.2	5.2 ± 0.1	1160 ± 121
A7**	0.5 – 2	1.25	13.7 ± 0.1	600 ± 31
A9**	0.5 – 2	1.25	17.0 ± 0.3	998 ± 17

* cork powder
** granulated industrial cork from the outer plank section

Each sample was extracted using a conventional Soxhlet solvent extraction system; Table 2 shows the parameters for the Soxhlet extraction: ca. 8 g were extracted with each one of the four solvents – ethanol, acetone, dichloromethane and hexane – for about 8 h. The solvent was removed by evaporation at reduced pressure and then under a nitrogen stream. The dried extracts were stored at –5°C. Extraction essays were performed in triplicate.

Table 2: Parameters for the Soxhlet extractions [6]

Samples	A1 up to A9
Solvents	Hexane (HEX), dichloromethane (DCM), acetone (ACE) or ethanol (EtOH)
Mass of sample	6 – 10 g
Ratio of solvent/solid matrix	12 – 20 mL$_{solvent}$/g$_{Cork}$
Extracion time	8 h

Results and Discussion

To evaluate the cork extractives composition, some studies [2-4] perform a sequential soxhlet extraction with several solvents of increasing polarity and the sum of this sequential extraction will represent the total cork extractives. Table 3 shows the yields reported for several cork samples: virgin cork, reproduction cork, cork powder and natural boiled cork; the total extractives were between 56 and 169 g$_{extract}$/kg$_{Cork}$. Note that in those studies all cork samples were granulated to particle sizes lower than 1 mm.

Table 3: Yields for sequential extractions of several cork samples.

Material	Total extractives [%]	DCM	EtOH[1] or MeOH[2]	H_2O	Ref.
Virgin Cork	16.9	7.9	5.8[1]	3.1	[2]
Industrial cork powder	5.9	2.6	1.9[2]	1.4	[4]
Reproduction Cork	16.2	5.8	5.9[1]	4.5	[3]
Reproduction Cork	14.2	5.4	4.8[1]	4	[2]
Natural boiled cork planks obtained after a cork cooking stage	8	4.7	2.2[2]	1.1	[4]
Inner from boiled cork planks obtained after a cork cooking stage	11.2	5	3.6[2]	2.6	[4]
Outer from boiled cork planks obtained after a cork cooking stage	5.6	2.7	1.8[2]	1.1	[4]

It is known that the cork composition depends on several factors, as growth conditions, genetic origin, age and tree dimension, etc. [1]. It is assumed as a benchmark that the total extractives for the samples used in this work, from the inner section of cork planks, is 112 $g_{extract}/kg_{DryCork}$, according to the value reported by Sousa et al. [4].

In the soxhlet extraction, for hexane, dichloromethane, acetone or ethanol, the maximum yield obtained, in this work, was 94.7 $g_{extract}/kg_{DryCork}$ for the more polar solvent (ethanol); most probably, the cork samples used, after the extraction, still contain some extractives, since no sequential extraction was conducted; and when the solvent polarity decreases, the fraction of extractives that remains in the sample is higher. Table 4 represents the maximum yield that was obtained for each solvent.

Table 4: Yields for single soxhelt extraction with different solvents.

Solvent	Maximum yield [$g_{extract}/kg_{Cork}$]
Hexane (HEX)	53.5
Dichloromethane (DCM)	70.4
Acetone (ACE)	84.3
Ethanol (EtOH)	96.4

The soxhlet extraction was conducted until completion; the maximum yield achieved for each solvent represents the capacity for the respective solvent.

When the amount extracted per kg of sample is plotted against the particle size as shown in Fig. 2, it is clearly seen that this specific amount decreases with the particle dimension. It is not reasonable to presume that the total amount of extractables vary with the particle size for samples from the same region of the plank. Then, it is assumed that only an outer layer of cork is extracted when the particle size increases. As it is well known the cork is an excellent liquid sealant used from ancient times until nowadays. It seems then reasonable to assume that any liquid in contact with cork will penetrate just its outer layer, being prevented to attain the inner core due to the unique cork sealant properties. Assuming that each solvent extracts its maximum from the wetted region, it is possible to estimate the outer layer as a function of the extracting solvent.

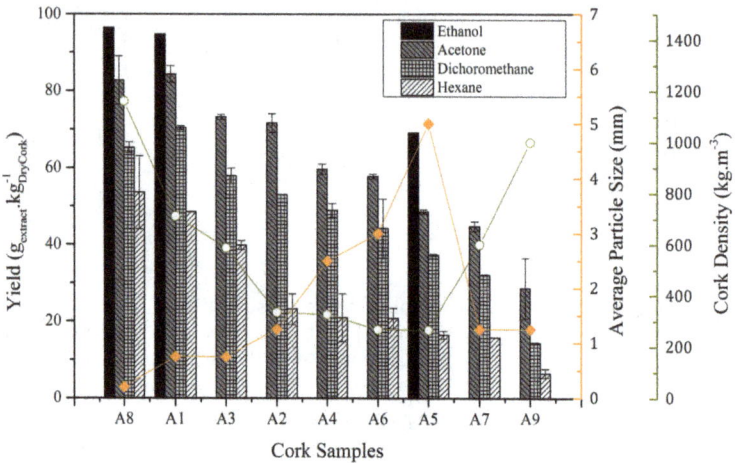

Fig. 2 Yields of conventional soxhlet extraction of industrial granulated cork (samples: A1 up to A9) as a function of particle size

To evaluate the cork layer accessible to be extracted from each sample, it was assumed that the extractives are equally distributed inside the particle of each sample, that all particles are spherical and also that cork powder was completely extracted, i.e., $\xi_c = r_c/R_p = 0$, where ξ_c represents the dimensionless radius, R_p is the particle radius and r_c the radius of core free of solvent. Fig. 3 shows when the particle it is totally extracted, yields achieved is the maximum.

Cork Science and its Applications (CSA2017)
Materials Research Proceedings **3** (2017) 32-39

Materials Research Forum LLC
doi: http://dx.doi.org/10.21741/9781945291418-5

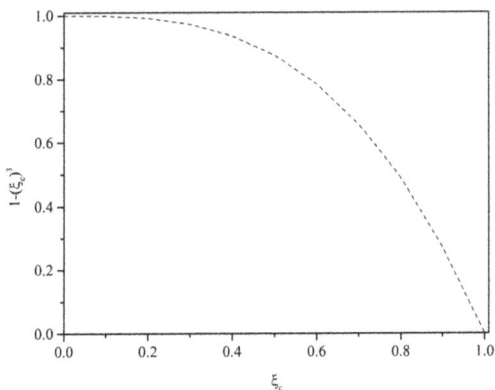

Fig. 3 Dimensionless volume of the extractable cork layer vs. dimensionless (non-extracted) core radius, ξ_c.

Based on the aforementioned assumptions, Fig. 4 displays the experimental values of the dimensionless volume of the extracted cork layer for each particle and solvent, as function of the particle size. When the particle size increases, the dimensionless thickness of the extracted (outer) layer decreases. It is possible to note that for particles bigger than 1 mm this value decreases only slightly; then, it can be considered that this layer is nearly constant for each solvent; being for hexane around half the value extracted with the acetone and dichloromethane, regardless of particle size. The dimensionless volume, on average, is ca. 0.22, 0.20 and 0.11 for acetone, dichloromethane and hexane, respectively [6]. For ethanol, it is only considered one particle size, and the dimensionless volume is 0.16.

Cork Science and its Applications (CSA2017)   Materials Research Forum LLC
Materials Research Proceedings **3** (2017) 32-39   doi: http://dx.doi.org/10.21741/9781945291418-5

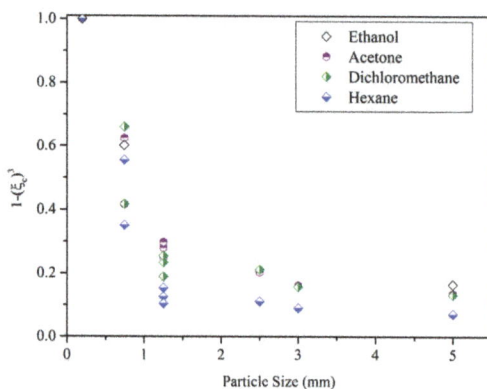

Fig. 4 Dimensionless volume of the extractable outer cork layer vs. average particle size.

With these results it is possible to evaluate the total extractives that can be removed using a specific solvent with particles of known size, as shown in Fig. 5; this information can be useful when the extracted cork samples are to be used in their traditional applications, after extraction of the triterpenic compounds. Also, knowing the thickness of the outer layer accessible to extraction is an essential information to be used when dealing with the modelling of the extraction process, a natural continuation of this work, to be pursued in the future.

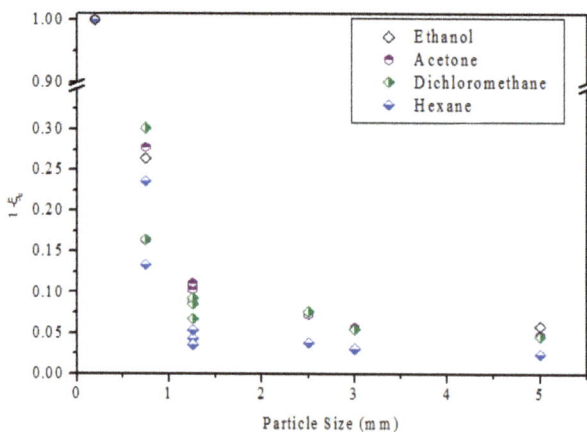

Fig. 5 Dimensionless thickness of the extractable outer cork layer vs. average particle size

Conclusions

In this work, the thickness of the outer layer of cork accessible to extraction for granulated industrial cork after the cooking stage was evaluated for several solvents. As cork is an excellent liquid sealant, it is to expect that any liquid in contact with cork will penetrate just its outer layer. Assuming that each solvent extracts its maximum from the wetted region, it was possible to estimate the outer layer as a function of the extracting solvent. The results show that, for particles bigger than 1 mm, the dimensionless volume of the extractable spherical crown is nearly constant and only depends on the type of solvent; on average, it is ca. 0.22, 0.20 and 0.11 for acetone, dichloromethane and hexane, respectively.

Acknowledgements

This work was financially supported by Fundação para a Ciência e Tecnologia (Portugal) through the PhD grant SFRH/BD/71891/2010. The work was also co-financed by: Project POCI-01-0145-FEDER-006984 – Associate Laboratory LSRE-LCM funded by FEDER through COMPETE2020 - Programa Operacional Competitividade e Internacionalizaçao (POCI) – and by national funds through FCT - Fundaçao para a Ciência e a Tecnologia.

References

[1] E. Conde, E. Cadahía, M.C. Garcia-Vallejo and J.R. Gonźalez-Adrados, Chemical Characterization of Reproduction Cork from Spanish Quercus Suber, Journal of Wood Chemistry and Technology, vol. 18, pp. 447–469, 1998. https://doi.org/10.1080/02773819809349592

[2] H. Pereira, Chemical composition and variability of cork from Quercus suber L, Wood Science and Technology, vol. 22, pp. 211–218,1988. https://doi.org/10.1007/BF00386015

[3] H. Pereira, Variability of the Chemical Composition of Cork, BioResources, vol. 8, no. 2, pp. 2246–2256, 2013. https://doi.org/10.15376/biores.8.2.2246-2256

[4] A.F. Sousa, P.C.R.O. Pinto, A.J.D. Silvestre and C.P. Neto, Triterpenic and other lipophilic components from industrial cork by products, Journal of Agricultural and Food Chemistry, 54, pp. 6888–6893, 2006. https://doi.org/10.1021/jf060987+

[5] APCOR. (2015). "Cork Yearbook 2015".

[6] Y.A. Manrique, Supercritical Fluid Extraction and Fractionation of Bioactive Natural Products from Cork, Ph.D. Thesis, Faculdade de Engenharia da Universidade do Porto (FEUP), Porto/Portugal 2017.

Cork Science and its Applications (CSA2017)   Materials Research Forum LLC
Materials Research Proceedings 3 (2017) 40-45   doi: http://dx.doi.org/10.21741/9781945291418-6

A Cost-Effective Methodology to Perform Customized Moulding of Cork Agglomerates

Daniel Afonso[1,2,a], Ricardo Alves de Sousa[1] and Ricardo Torcato[2,3]

[1] TEMA: Centre for Mechanical Technology and Automation, Department of Mechanical Engineering, University of Aveiro, Campus de Santiago 3810-183 Aveiro, Portugal

[2] School of Design, Management and Production Technologies Northern Aveiro, University of Aveiro, Oliveira de Azeméis, Portugal

[3] CICECO: Centre for Research in Ceramic and Composite Materials, University of Aveiro, Campus de Santiago 3810-183 Aveiro, Portugal

[a]dem.secretaria@ua.pt, [b]esan.geral@ua.pt, [c]ciceco@ua.pt

Keywords: Compression Moulding, Cork, Rapid Tooling, SPIF

Abstract. The use of cork–polymer composites allows great design possibilities, combining the advantages of the natural material and melt based technologies to achieve unique part properties. An attractive processing technique is compression moulding for being a free form low waste process, and allowing the recyclability of other cork products waste. Still, the process requires the development of medium to high complexity tools, restricting the process to medium to high production volumes. This research tests the development of low cost solutions for the development of cork compression moulding tools using sheet metal. Single point incremental forming is used as a rapid tooling process for the manufacture of cork compression moulding tools. A simple part is used as a case study example and the mould is design and build while defining general guidelines for sheet metal low cost tooling. The developed tool is used in two steps moulding operation by compressing and locking the mould before heat curing the resin, proper for a low batch manufacture process. The tool integrity is checked and the moulded part is evaluated for quality before final considerations are set.

Introduction

One of the best ways to introduce innovation in a business context is by turning innovation into something meaningful that responds to real consumer needs. This may be achieved by planning new products where design becomes the driving force in successful new markets. Taking into account the international decrease in the market for cork stoppers and the great potential of cork materials and technology, the cork industry search for new commercially successful market solutions [1]. As a possible response, the use of cork finds great applicability in the development of sustainable products. For that, different material cork materials and processing technologies may be used [2, 3], where cork compression moulding is one of the most free form capable and low waste techniques.

The cork compression moulding is suitable for processing cork–polymer composite (CPC) in order to develop products with new shapes from theses that cork can provide for acoustic, thermal insulation, energy absorbing or aesthetic applications [4]. This process typically uses a composite of 0.5 to 1.0 mm grain size cork granules, possibly recovered from other cork processing technologies waste, and a binder resin, either thermoset or thermoplastic. The granulated cork density varies from 60 to 160 kg/m^3 and can be agglomerated to 140 to 600 kg/m^3 by compression moulding operations. Due to the absence of waste material, this manufacturing process allows to achieve fairly complex geometries for medium to high quantities. [2,4]

Cork Science and its Applications (CSA2017)   Materials Research Forum LLC
Materials Research Proceedings 3 (2017) 40-45   doi: http://dx.doi.org/10.21741/9781945291418-6

A relevant consideration for the use of compression moulding for a given part design due the tooling costs, inhibiting its use for low batch manufacture. The tool development requires purpose built tools a part design. Depending on the intended moulded part density, the moulding load may varies from 0.5 to 2.0 MPa. When operation with compression ratios bellow 2, typically aiming for density bellow 200 kg/m^3, the moulding load is typically inferior to 0.6 to 0.7 MPa. [3,5] Since this moderate pressure is used in cork compression moulding, it is possible to replace conventional tooling materials and design by sheet metal based tools.

Single point incremental forming (SPIF) is a smart manufacture process, capable of forming sheet metal is a free form geometry. SPIF can be used to shape sheet metal to the desired surface without the need for purpose built tools, thus being suited for unique part manufacture. Apart from be a potential faster and more economical tooling process, a low thermal inertia may benefit the mould operation has it could speed up the heating time. Nevertheless, the fabrication and the mechanical behaviour of the sheet metal mould as some geometry related issues.

The definition of a SPIF part could fallow different configuration, where a container shape is the most common. During the forming operation, thinning occurs, where the final thickness can be estimated by the sine-law ($t_f = t_i \times sin(90-\alpha)$), where α is the wall angle. The parts can be made from multiple materials, mainly metals with thickness up to 3 mm. The achievable geometry is primarily limited by a maximum wall angle depending on the material, thickness and forming conditions and machine operation limits in size and strength [6].

SPIF Sheet Metal Mould Design
A generic compression mould tool is defined by the complementary surface of the part. Typically, tools uses a two part assembly with a cavity and a core side. Parts thickness can vary from only 1 mm to over 25 mm, allowing to shape thicknesses variations along the part. The compression moulding part require the use of a minimum draft angle from 3° and a minimum 2 mm radius in all edges. The process allows the shaping of free form surfaces, allowing the inclusion of corrugated features, bosses or ribs. [7] In cork compression similar guidelines are taken, with special concern to minimum thickness to grant part integrity and minimum detail sized according to the used granulates. The definition of large solid volumes is also possible.

The test part is designed as a five sides box with a minimum draft angle of 20° and a wall thickness between 26 and 37 mm. The tool for the cork compression moulding SPIF rapid tooling research uses a two pieces mould with a flat parting line. The mould is design for a vertical operation, with the part being moulded upside down. One part of the mould is formed by a cavity with the complementary surface of the upper side of the designed part. The opposite part is formed by a core with the complementary geometry of the part hollow to mould the thick wall part. This part is sized using the sine-law to predict the sheet metal thickness after forming due to thinning. This two parts are the moulding essentials, designed as sheet metal containers. Since the raw material volume is much higher than the empty cavity, the mould assembly is completed with a frame box to allow the placing of all the uncompressed cork mixture. This frame box is defined to be manufactured by a flange type SPIF part with a manually bent extension up to 40 mm height to allow the deposition of sufficient material to reach a compression ratio of 2. Finally, for a more convenient operation, a support box in included in the mould design. Besides, a clamping system is added to allow locking the closed mould under pressure for oven resin curing. Figure 1 presents the sheet metal mould concept for cork compression moulding operation. The presented mould concept not only responds to the designed part but also fallows a common guideline for any flat parting line part. Moulds with a higher cavity volume over part volume ratio may dispatch the frame box extension. Compression moulding tools for other materials where the material compression ratio is lower may also dismiss the frame box.

Cork Science and its Applications (CSA2017)   Materials Research Forum LLC
Materials Research Proceedings 3 (2017) 40-45   doi: http://dx.doi.org/10.21741/9781945291418-6

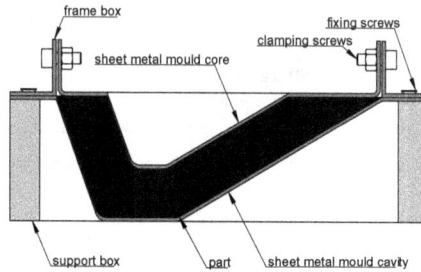

Fig. 1 Cork compression moulding sheet metal mould concept

In what concerns the mechanical behaviour of the tool must be sized to support the moulding pressure. The support box is dimensioned so that the cavity sheet metal part is both supported around the perimeter an in the bottom. The sheet metal parts initial thickness is sized to hold out the moulding pressure. The mechanical behaviour is tested by numerical analysis to support distributed loads up to 0.2 MPa. The mould is defined using a 2 mm thickness 1050-H111 aluminium alloy. In order to support higher loads, two option can be carried out. One the one hand, a thicker sheet or stiffer material can be used for the mould design. On the other hand, additional support can be added to the mould, either by defining a grid of reinforcements or filling the mould with sand or other porous mixture. [8]

Tests and Validation

The mould manufacturing is performed on the SPIF-A machine. The forming operation of the three SPIF parts is done using a 12 mm spherical punch with a 0.5 mm vertical forming step in a helical tool path strategy. The container configuration parts are formed in a single stage tool path strategy and the flange configuration part is formed using a multi stage strategy.

After SPIF forming operations, complementary mould handiwork is performed. The frame box extension is applied by hand bending and riveting a 1 mm thickness aluminium strip. The core side part is cut and bent to fit in the frame box with minimum gap. All parts are drilled and fixed in position on the support box. Figure 2 presents the complete open sheet metal compression mould.

Fig. 2 Cork compression mould sheet metal tool

Cork Science and its Applications (CSA2017)   Materials Research Forum LLC
Materials Research Proceedings **3** (2017) 40-45   doi: http://dx.doi.org/10.21741/9781945291418-6

The total manufacture process is performed in under 4 hours and an half, being 56% of the time for the SPIF process itself. The total energy consumption is 17.2 kW.h and the total material cost is 20.00€, including aluminium sheets, MDF support box and forming backing plate. The mould parts are measured, registering a deviation of ±6 mm and an average deviation of +1 mm on both the cavity and the core parts.

Aiming for a 180 kg/m^3 density, the part weight is 167 g. Considering a 9:1 weight ratio, the CPC uses 150 g of cork powder and 17 g of thermoset PU bounding resin. In order to add some moisture to the powder mixture, 10 g of water are added. The cork compression moulding operation is performed in a two steps operation, first compressing the cork mixture and then curing in an oven. A silicone based demoulding agent is used and demoulding paper pieces are positioned in plannable areas to help the part release. A 15 tones hand actuated hydraulic press is used is used for the material compression. The maximum applied force reach 500 kg.f, corresponding to a moulding pressure of 0.15 MPa considering the 180 × 180 mm projected area. During the mould compression, a small amount of material is lost by the side locking holes. Yet the amount can be despised when compared with the moulding material volume. The mould is locked while under pressure using the side screws. The frame box support the closed mould under pressure without core part slide as the press force is released. The closed mould is removed from the press and cured on an oven for two hours with a temperature set for 140 °C with maximum peak of 150 °C. After cured, the part is released with ease from the mould.

After demoulding operation the part is cleaned and the non-agglomerated material around the bottom edge is removed.

Given the two operation steps, the moulding operation is performed under two and an half hours, including the cork mixture preparation, compression and cure. Together with the mould manufacture process, this allows the achievement of a first part within a day of work, reaching the goals of a rapid tooling system. Figure 3 presents the moulding compression operation and the demoulded part.

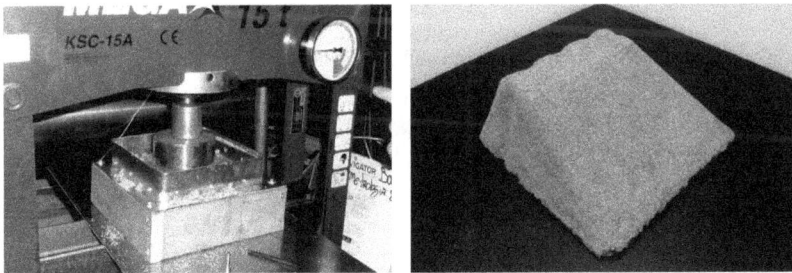

Fig. 3 Cork compression moulding operation: (left) mould under pressure, (right) demoulded part

The part is first evaluated trough visual inspection. The evaluation assesses the mould filling, part apparent density and general part quality. The part success to define the overall geometry definition, with only minor failures at the sharp edges, small defects due to wrinkles in the demoulding paper at the bottom side of the part and some high concentration binder spots noticeable at surface. Nevertheless, the surface quality is good and cork granulated agglomeration apparent distribution is uniform.

The part is weight for density evaluation. The total part weight is 166 g, resulting on and effective global density of 180 kg/m^3. During the moulding process the material lost is negligible. A manual compression of the part surface suggest an uniform material behaviour, an so a uniform material density.

The part is measured for accuracy evaluation. Despite some significant deviations, the cork part is moulded with an average deviation of −0.4 mm. However, the largest deviation along the part surface goes up to 6.3 mm at the centre of the lower slope wall, due to the mould deformation. A biggest deviation is found at the lower edges with the real part 13 mm smaller than the CAD model. This large value is partly due to the failure in agglomerating the material at the part bottom edge because of gap between the core part and the frame box and sharp edge between the upper and lower side of the part.

After operation, the mould parts are measured to check for permanent deformation. The core side of the mould suffered a permanent deformation close to 3.5 mm at the centre of the edge more far from the core feature. Besides, a mark from the pressure distribution plate of the press in noticeable at the part. The cavity side was deformed by 4.0 mm at the centre of the lower slope wall. All remaining walls suffer no permanent deformation.

Conclusion

The use of sheet metal moulds for cork compression moulding operation is validated. Some relevant issues are still found at the moulded parts. Nevertheless, some of these problems, particularly the binder spots and the wrinkle marks result from the compression moulding operation and not to the mould itself. Only the miss definition of the part edge is affected by the mould, due to the considerable gap between the frame box and the core side. Even so, the use of sheet metal moulds is an attractive solution for compression moulding with cork or other powder and binder based composites, mainly when the production volume requires low cost tooling. In such a way, although some improvement must be done, SPIF is validated as a rapid tooling process for compression moulding.

The mould fabrication has reduced time and cost, and has the potential to achieve relatively complex parts. The inclusion of sheet metal parts with other parts increases the geometry limits and, in specific situations, lead to a faster mould development and lower tooling cost.

From the mechanical point of view, the 2 mm sheet metal alone is sufficiently rigid although not tough enough to support the moulding loads, leading to permanent deformation. On the core side of the mould, the permanent deformation result from an uneven pressure distribution. The addition of a correct size pressure plate could eliminate this issue. On the cavity side, the deformation dues to the geometry itself. The mechanical behavior could be improved by either use a thicker sheet or by adding a filling support by using sand or other porous mixture.

Apart from the defects that result from the mould deformation and oversized gap between the core and the frame box and operation issues, the part quality is good. The unaffected surfaces have an accuracy of ±2 mm and a good surface finishing. The punch marks on the sheet metal surface are not visible on the moulded part.

References

[1] A. Mestre and L. Gil, "Cork for sustainable product design," Ciência & Tecnologia dos Materiais, Vol. 23, n.º 3/4, 2011.

[2] Amorim Cork Composites, "Cork Solutions & Manufacturing Processes," 2017.

[3] E. Fernandes, V. Silva, J. Chagas, and R. Reis, "Cork-polymer composite (cpc) materials and processes to obtain the same", WO Patent App. PCT/PT2008/000,051, 2009.

Cork Science and its Applications (CSA2017)   Materials Research Forum LLC
Materials Research Proceedings **3** (2017) 40-45   doi: http://dx.doi.org/10.21741/9781945291418-6

[4] E. Fernandes, V. Correlo, J. Chagas, J. Mano and R. Reis, "Cork based composites using polyolefin's as matrix: Morphology and mechanical performance," Composites Science and Technology, Vol. 70, n° 16, 2010. https://doi.org/10.1016/j.compscitech.2010.09.010

[5] S. Silva, M. Sabino, E. Fernandes,V. Correlo, L. Boesel and R.Reis, "Cork: properties, capabilities and applications," International Materials Reviews, Vol. 50, n° 6, 2005. https://doi.org/10.1179/174328005X41168

[6] D. Afonso, R. Alves de Sousa, and R. Torcato, "Defining design guidelines for single point incremental forming," Sustainable Smart Manufacturing conference, 2016.

[7] Molded Fiber Glass Companies, "Technical Design Guide for FRP Composite Products and Parts - Techniques & Technologies for Cost Effectiveness", 2016.

[8] R. Appermont, B. Van Mieghem, A. Van Bael, J. Bens, J. Ivens, H. Vanhove, A. Behera, and J. Duflou, "Sheet-metal based molds for low-pressure processing of thermoplastics," Proceedings of the 5th Bi-Annual PMI Conference, 2012.

Cork Science and its Applications (CSA2017)   Materials Research Forum LLC
Materials Research Proceedings 3 (2017) 46-59   doi: http://dx.doi.org/10.21741/9781945291418-7

Is Cork a Good Closure for Virgin Olive Oil?

Ofélia Anjos[1,2 a *], Luís Coutinho[1,b], Cecília Gouveia[1,c] and Fátima Peres [1,3,d]

[1] Instituto Politécnico de Castelo Branco, Castelo Branco, Portugal

[2] Centro de Estudos Florestais, Instituto Superior de Agronomia, Universidade de Lisboa, Lisboa, Portugal.

[3] LEAF, Instituto Superior de Agronomia, Universidade de Lisboa, Lisboa, Portugal.

[a]*ofelia@ipcb.pt, [b] tojeiraprodutosbiologicos@gmail.com, [c] cgouveia@ipcb.pt, [d] fperes@ipcb.pt

Keywords: Cork Stopper; Screw Cap; Storage; Virgin Olive Oil

Abstract. Most of the studies on olive oil preservation during storage are focused on the type of conditioning rather than on the most efficient type of bottle seal to be used. However, the bottle closure is also an important issue because of the negative impact that oxygen has on olive oil quality and flavor.

The aim of this study was to assess the performance of a natural cork stopper as a closing system of glass bottles for olive oil.

To evaluate the effect of a bottle closure, a storage trial over 24 months, after harvest, was performed comparing the effect of three types of glass bottle closure on virgin olive oil quality: screw cap, natural cork and bar top cork stopper. The bottle neck of those with a natural cork stopper was covered with bee's wax. The results for quality criteria showed that all three types of bottle closure acted in a very similar way for most of the studied parameters. However, based on FTIR-ATR spectral information it was possible to separate the samples with different closure systems at the end of the storage period. These differences could be given by the retention of some volatile compounds detected by sensory evaluation.

The olive oil oxidation parameters were not highly affected by the cork stopper. However, for olive oil quality characterization with a cork stopper, more studies are needed for the volatile composition.

Introduction

The cork oak forests have great relevance for Portuguese industry and for the landscape, given their ecological importance against desertification caused by the variety of animal and plant biodiversity in their stands.

The cork oak stands grow in several western Mediterranean countries, with a total worldwide area of 2.3 million hectares. Portugal is one of the most important producer and transformers of cork with around 34 % (736 thousand hectares) which corresponds to about 23 % of the national forest. As for cork production, Portugal has 49.6 % of the total world production.

Cork is the best material for sealing wine bottles, however as far as we know it is not used extensively to seal other food products.

Cork is the outer layer on the cork oak tree bark (*Quercus suber* L.) and it is usually harvest in intervals of 9 years. It is a low density cellular material with great compressibility and dimensional recovery [1,2], has appropriate mechanical properties [3–6], insulation properties, very low permeability to liquids and gases, and chemical stability and durability [7,8].

The anatomical characteristic of cork that give it these mentioned properties is that it is formed by closed cells that contain air-filled lumens, lenticular channels and other tissue with very small air-filled pockets [6]. Given this characteristic and consequently a cork stopper, has low permeability to oxygen [9,10]. However, the cork stopper is not perfectly impermeable, which is an important characteristic for wine. In this case for wine sealing, the oxygen transmission rate into the closed bottle is an important parameter for the wine cellars, given its relation to the quality development of the wines [11–16]. Nevertheless, this characteristic so important for wine quality could be a disadvantage for other uses of cork as a stopper.

Virgin olive oil is extracted from the fruits of the olive tree (*Olea europaea* L. var. *europaea*) using exclusively mechanical means. The fact that the oil extraction is solvent-free and natural antioxidants which are maintained in the oil is reflected in the nutritional and sensory properties as well as in the economical value of this product. So, during the storage of olive oil a continuous effort should be made in order to preserve its high quality features.

The main deteriorative reaction that occurs during storage is oxidation. Many studies have been published about the negative impact that oxygen has on olive oil, namely promoting the formation and degradation of hydroperoxides, originating volatile compounds of low molecular weight responsible for the degradation of the sensory profile usually called rancidity [17–20].

Cloudy, or veiled VOO contains phenols, phospholipids and sugars, but it can also contain hydrolytic and oxidative enzymes, such as lipases, lipoxygenases, and polyphenol oxidases. These enzymes may reduce the "pungent" and "bitter" sensory notes, the intensity of which is strictly linked to the content of aglycon secoiridoids, and, at the same time, can produce olfactory and taste defects [21]. Consequently, the olive oil profile changes during its storage due to the simultaneous drastic reduction in compounds formed by the LOX pathway and to the formation of new volatile compounds responsible for some common defects such as "rancid", "cucumber" and "muddy" [22–24]. This is accompanied by the increase in saturated aldehydes nonanal and hexanal, coming from the oxidation process [22].

The studies conducted on olive oil storage have mainly been focused on the type of conditioning [25,26]. As a matter of fact packaging materials that have been tested (clear glass, clear polyethylene terephthalate (PET), clear PET+UV blocker, clear PET covered with aluminum foil and clear polypropylene (PP) bottles) have shown that the best packaging material for olive oil was glass followed by PET [27]. Those studies concluded that exposure of olive oil samples to light, high storage temperatures (35 °C) and large headspace volumes caused substantial deterioration in product quality parameters.

As far as we know, the type of closure used for olive oil glass bottle was never researched. However, the type of closure can have a substantial impact on the chemical and sensory composition of olive oil. The present paper intends to continue research comparing screw cap bottle closure with cork stoppers.

Material and methods
Samples and experimental design. Virgin olive oil used in this study was produced in 2014 from olives from the two main Portuguese cultivars: 'Galega vulgar' and 'Cobrançosa' blended in the proportion 1:1 (v/v). The olive oil was produced in the "Beira Interior", and can be labelled as PDO (Protected Designation of Origin) and belong to the olive oil awarded at the Mario Solinas Competition in the category of fruity mature. The olive oils were packaged in 36 dark glass bottles with a volume of 0.5 L (Figure 1) and sealed with three different types of closures:
- 12 bottles with screw caps (SC);
- 12 bottles with cork stopper lined with bee's wax (CW);
- 12 bottles with Bartop natural cork stopper (Bt).

Cork Science and its Applications (CSA2017)   Materials Research Forum LLC
Materials Research Proceedings 3 (2017) 46-59   doi: http://dx.doi.org/10.21741/9781945291418-7

Fig. 1 *Experimental units scheme.*

Table 1 – Analytical methods used for olive oil characterization and for evaluating the changes during the storage period.

Parameter	Method	Reference
Acidity (% oleic acid)	Volumetric analysis (acid/base)	[28] Annex II
Peroxide value (meq O_2 kg^{-1})	Volumetric analysis (redox)	[28] Annex III
UV absorbances	UV spectroscopy	[28] Annex IX
Organoleptic assesment	Trained assessors	[28] Annex XII
Fatty acid composition (%)	GC-FID	[28] Annex XA
Triacylglycerides (%)	HPLC-RID	[28] Annex XVIII
Total phenols (mg GAE kg^{-1})	VIS spectroscopy	[29]
Tocopherols (mg kg^{-1})	HPLC-FLD	[30]
Chlorophyll pigments (mg kg^{-1})	VIS spectroscopy	[31]
K_{225}	UV spectroscopy	[32]

Analytical methods. To evaluate the effect of the bottle closure on the olive oil quality over time, oils were tested after 6, 18 and 24 months storage (room temperature of 22 °C). At each of these analysis periods, two bottles of the 36 bottles originally stored for each storage condition was removed from storage and the oil was analyzed. That oil was then discarded and not used again in the study. These time intervals were selected given previous studies performed by Peres et al. [33] that showed that the major changes in olive oil oxidation were after the first six months of storage, using plastic closures.

Cork Science and its Applications (CSA2017)   Materials Research Forum LLC
Materials Research Proceedings **3** (2017) 46-59   doi: http://dx.doi.org/10.21741/9781945291418-7

FTIR-ATR data acquisition. The olive oil FTIR-ATR spectra were acquired in Bruker FTIR spectrometer (Alpha) with a resolution of 4 cm^{-1} in the wavelength region of 4000-400 cm^{-1}, using a diamond single reflection attenuated total reflectance (ATR). All spectra were obtained with 32 scans and the experiments were carried out at room temperature. The background measurement was done using air.

To identify the differences between samples a PCA was performed with the spectral data and some pre-processing tools were applied to the raw data: multiplicative scatter correction (MSC); minimum maximum normalization (MinMax); vector normalization (VecNor); straight line subtraction (SLS); constant offset elimination (ConOff); first derivative (1stDer); second derivative (2ndDer); first derivation with multiplicative scattering correction (1stDer + MSC); first derivation with vector normalization (1stDer+VecNor); first derivation with straight line subtraction (1stDer + SLS).

Statistical analysis. A factorial variance analysis was performed to assess the effects of the closure type and storage time as fixed factors. The factorial design was performed with two levels. For each significant effect or interaction, the variance percentage was calculated and a Scheféé post-hoc test for a significance level of $\alpha = 0.05$, was applied to the corresponding variables.

For the ANOVA results were considered the different levels of significance as a function of the p value: *** - highly significant for $p < 0.001$; ** - very significant for $p < 0.01$; * - significant for $p < 0.05$; n.s. - not significant for $p > 0.05$.

The results were also subjected to a multivariate analysis (Principal Component Analysis) to compare the similarity between samples. All the calculations were performed using Statistic (vs 7.09) from StatSoft and Minitab® 17.1.0.

Results and discussion

The triacylglycerol composition of the blend of olive oil used in the present study, based on the equivalent carbon number (ECN) was the following: ECN 42=0.2; ECN44=3.5; ECN 46=20.8; ECN 48=6.3 and ECN50= 6.0. The result of triacylglycerol composition and consequently the fatty acid composition (Table 2) shows that the blend is characterised by a high content of MUFA and a low content of PUFA, which has already been said by Peres et al. [30] for those olive oils that contribute to the blend.Virgin olive oils are known to be more resistant to oxidation than other edible oils because of their natural antioxidant content, particularly phenolic compounds and relatively low content of PUFA. Extra virgin olive oil (EVOO) is the highest grade of olive oil and to be classified as EVOO the oil must contain less than 0.8 % free fatty acids (measured as oleic acid) and the peroxide content must not exceed 20 meq O_2/kg of oil, as well as many other components such as UV absorption and organoleptic assessment. The results obtained showed that the blend has: 0.3% acidity (Table 2), 6 meq O_2/kg of peroxide index, specific absorbencies at 270 nm equal to 0.15 and at 232 nm of 1.6 (Table 3). In what concerns the organoleptic assessment the results showed no defects and a fruity flavour of 6. Alpha-tocopherol, the most important tocopherol in olive oil, was also evaluated in the beginning of the trial (384.25 mg/kg) showing similar contents with [21].

Although the storage period studied was after 6, 18 and 24 months, the fatty acid composition was only evaluated after the period that corresponds to the legislation (18 months storage). As can be observed (Table 2) fatty acid composition of the oil was not influenced by the storage in different types of closure. After 18 months, free acidity was also evaluated and the results showed an increase of 0.1% after the storage period but all the oils had similar behaviour (Table 2). The increase of 0.1% after 18 storage was already been referred by other authors [33].

Tabela 2 – Fatty acid composition (%) and acidity (% oleic acid) in the olive oil at the begining of the trial (T_0) and after 18 months after harvest (T_{18}) (Mean ± standard deviation).

Fatty acids	Início (T_0)	SC(T_{18})	CW (T_{18})	Bt (T_{18})
Miristic acid (C14:0)	0.01 ± 0.00	0.01 ± 0.00	0.01 ± 0.00	0.01 ± 0.00
Palmitic acid (C16:0)	14.56 ± 0.13	14.61 ± 0.04	14.66± 0.03	14.64 ± 0.01
Palmitoleic acid (C16:1)	1.73 ± 0.01	1.75 ±0.01	1.74 ±0.00	1.74 ±0.00
Margaric acid (C17:0)	0.12 ± 0.00	0.12 ±0.00	0.12 ±0.00	0.12 ±0.00
Margaroleic acid C17:1	0.24 ± 0.00	0.24 ±0.00	0.24 ±0.00	0.24 ±0.00
Stearic acid (C18:0)	3.17 ± 0.25	3.45 ±0.28	3.16 ±0.00	3.15 ±0.00
Oleic acid (C18:1)	70.88 ± 0.06	70.67 ±0.31	71.13 ±0.04	70.97 ±0.27
Linoleic acid C18:2)	7.54 ± 0.02	7.43 ±0.02	7.49 ±0.00	7.47 ±0.25
Linolenic acid (C18:3)	0.85 ± 0.00	0.84 ±0.00	0.85 ±0.02	0.85 ±0.01
Arachidic acid(C20:0)	0.43 ± 0.00	0.42 ±0.00	0.41 ± 0.00	0.42 ±0.00
Gadoleic acid (C20:1)	0.24 ± 0.00	0.24 ±0.00	0.24 ±0.01	0.24 ±0.00
Behenic acid (C22:0)	0.13 ± 0.01	0.13± 0.02	0.11 ±0.00	0.11 ±0.00
Erucic acid (C22:1)	0.00 ± 0.00	0.00 ± 0.00	0.00 ±0.00	0.00 ±0.00
Lignoceric acid (C24:0)	0.06 ± 0.00	0.05 ± 0.00	0.05 ±0.00	0.05 ±0.00
MUFA	73.19 ± 0.01	72.96 ± 0.30	73.41 ±0.00	73.25 ±0.26
PUFA	8.39 ± 0.02	8.26 ± 0.03	8.34 ±0.02	8.32 ±0.01
SFA	18.41 ± 0.12	18.78 ± 0.33	18.25 ±0.02	18.44±0.27
Free acidity (% oleic acid)	0.27 ± 0.00	0.41 ± 0.00	0.37 ± 0.00	0.38 ± 0.00

Chlorophyll is one of the main contributors to the color of virgin olive oils. It has been demonstrated that high concentrations of chlorophylls compromise the resistance to oxidation of olive oils exposed to light. Photo-oxidation of oil in the presence of chlorophylls leads to the formation of highly unstable and reactive singlet oxygen that tends to react with the unsaturated fatty acids leading to the formation of hydroperoxides [34].

Chlorophyll pigments in the present virgin olive oil corresponds to high contents [30] and as the storage was in the dark it can be observed that low changes are observed after 24 months of storage, not significantly different in olive oil with bartop closure (Table 3). However, the variation observed for chlorophyll pigments are mainly due by the storage period that is responsible for 91.4% of the total variance observed (Table 4).

Phenolic compounds are important minor components in olive oil which, due to the powerful antioxidant effect, contribute to the shelf life stability of olive oil [35]. The contents of total phenols in olive oil can have a huge range because of the dependence on cultivar, agronomic factors and storage conditions. In the present study total phenols in the oils corresponds to a medium content of phenols, but at the end of the storage period total phenols are still similar at the beginning of the trial, and with bartop closure no significant differences in phenolic content were observed (Figure 3). For both the chlorophyll pigments and for the phenolic compounds the storage period is the most significant factor and explains 79% of the total variance (Table 4). This trend is also verified for the K_{232} and K_{225} indexes.

The specific absorbance at 225 nm is an objective way to evaluate virgin olive oil's bitter taste and is highly correlated with total phenols [32]. In accordance with total phenols there were no specific trend of K_{225} and no significant differences were observed between time and type of closure. This is not in agreement with the results of Peres et al. [33], that showed a decrease in the first 6 months storage for specific absorbances at 225 nm.

Tabela 3 – Results of analytical parameters evaluated in olive oil during the storage period.

Parameter	0 months		6 months	18 months	24 months	Legal limits
Chlorophyll pigments (mg kg^{-1})	40.71 ± 0.01c	SC	40.01±0.04^b$_A$	38.87±0.18^b$_A$	37.79±0.07^a$_B$	
		CV	40.42±0.02b^c$_B$	39.82±0.49^b$_A$	38.20±0.34^a$_B$	
		Bt	40.01±0.06^b$_A$	40.17±0.08^b$_A$	38.37±0.20^a$_A$	
Peroxide index (meq O$_2$/kg)	5.99 ± 0.01a	SC	7.32±0.16^b$_B$	5.59±0.14^a$_A$	5.84±0.44^a$_A$	≤20.00
		CV	6.58±0.27^a$_A$	6.15±0.31^a$_A$	6.32±0.54^a$_A$	
		Bt	8.02±0.50^b$_C$	9.57±0.70^c$_B$	10.28±0.20^c$_B$	
K$_{270}$	0.15 ± 0.00a	SC	0.19±0.00^b$_{AB}$	0.23±0.00^d$_C$	0.21±0.00^c$_A$	≤0.22
		CV	0.19±0.00^b$_A$	0.16±0.01^a$_A$	0.22±0.01^c$_A$	
		Bt	0.21±0.04^b$_B$	0.20±0.00^c$_B$	0.21±0.01^d$_A$	
K$_{232}$	bc 1.60 ± 0.01 c b	SC	1.66±0.12^c$_A$	1.48±0.07^{ab}$_A$	1.39±0.03^a$_A$	≤2.50
		CV	1.60±0.01^c$_A$	1.51±0.02^b$_A$	1.37±0.01^a$_A$	
		Bt	1.68±0.06^c$_A$	1.61±0.01^b$_B$	1.48±0.01^a$_B$	
Total phenols (mg GAE/kg)	302.95 ± 4.64b	SC	308.4±6.5^b$_B$	255.8±2.2^a$_B$	250.1±21.9^a$_A$	
		CV	309.2±3.9^b$_B$	262.0±4.3^c$_B$	269.8±18.4^a$_A$	
		Bt	292.2±4.2^b$_A$	243.2±2.2^a$_A$	243.8±9.9^a$_A$	
K$_{225}$	0.20 ± 0.00ab	SC	0.20±0.00^a$_A$	0.20±0.00^a$_B$	0.21±0.01^b$_A$	
		CV	0.20±0.00^b$_A$	0.18±0.01^a$_A$	0.20±0.02^b$_A$	
		Bt	0.20±0.00^a$_A$	0.19±0.00^a$_B$	0.20±0.01^a$_A$	

Cork Science and its Applications (CSA2017)
Materials Research Proceedings 3 (2017) 46-59

Materials Research Forum LLC
doi: http://dx.doi.org/10.21741/9781945291418-7

Table 4 – ANOVA results for the quality parameters of olive oil.

Parameters	Variation origin (Variance percentage/significance level)			
	Closure (C)	months (M)	CxM	Residual
Chlorophyll pigments (mg kg^{-1})	1.9 0.0002(***)	91.4 0.0000(***)	3.0 0.0006(***)	3.6
Peroxide index (meq O$_2$/kg)	70.5 0.0000(***)	0.6 0.0255(*)	25.0 0.0000(***)	3.9
K$_{270}$	19.8 0.0000(***)	25.1 0.0000(***)	51.2 0.0000(***)	3.9
K$_{232}$	12.0 0.0000(***)	73.7 0.0000(***)	0.0 0.1065(ns)	14.3
Total phenols (mg GAE/kg)	9.9 0.0000(***)	79.0 0.0000(***)	0.0 0.2985(ns)	11.1
K$_{225}$	9.6 0.0150(*)	42.5 0.0000(***)	0.0 0.1647(ns)	47.9

*** highly significant $p<0,001$; ** very significant $p<0,01$; *significant $p<0.05$; n.s. – not significant $p>0.05$

For the parameters related to the olive oil oxidation (Peroxide index and K$_{270}$) the values of the interaction between closure system and months (CxM on Table 4) are highly significant and explain 25.0 % and 51.2 % respectively. These results are explained by the differences observed for the olive oil sealed with a bartop natural cork stopper during the storage period compared to the other closures; for K$_{270}$ parameter to the differences observed in all the closure systems during the storage period (Figure 2).

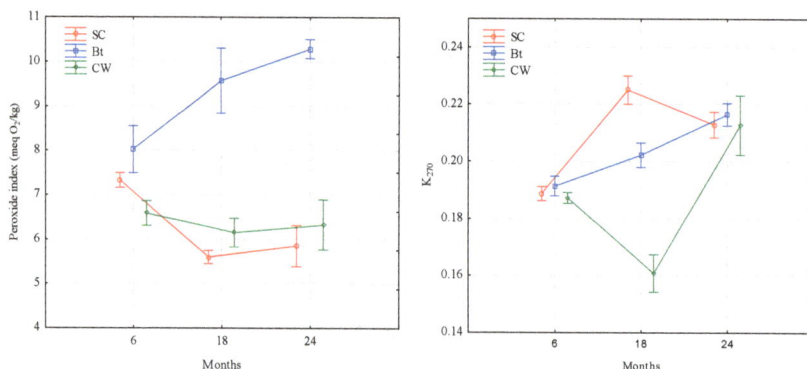

Fig. 2 Representation of the ANOVA interaction (CxM) for the peroxide index and K$_{270}$ parameters.

Figure 3 represents the results of the principal components analysis which explains 67% of the total variation. Component 1 separates the samples by the storage time with the exception of the sample that was sealed with cork and bee's wax for 18 months of storage. Component 2 separates the different closures tested in this study. It was observed that olive oils in bottles sealed with screw caps and cork stopper bee's wax have a similar behavior in terms of olive oil oxidation parameters and the olive oil in the bottle sealed with a bar

top natural cork stopper presented a different behavior related to the higher values of peroxide index and K_{270}.

A very interesting result for the cork industry is that the olive oil sealed with a cork stopper with bee's wax for 18 months has a very similar composition to those presented for other closures tested and is more similar to the olive oil composition at the beginning of the test.

The differentiation of the oil samples sealed with different closures at the end of the storage time in analysis is mainly due to the higher values of K_{270} and peroxide index, as shown previously with ANOVA results. The separation between the olive oil sealed with a bartop natural cork stopper and the other closures is mainly due by the K_{225} values. The authors think that the differences observed for the performance of the bartop natural cork stopper are given due to the possibility that air could be pass with this system.

It is well known that cork stoppers have low permeability to oxygen [9,10], which means that they are not perfectly impermeable. This characteristic is very important for wines but could be a problem for olive oil oxidation parameters. In the closure system with cork stopper and bee's wax this problem is solved with the bee's wax, but in the system with a bartop stopper it may not be possible to solve this. However, these points as well as the interaction between stopper and volatile composition of olive oil must be well researched in following studies.

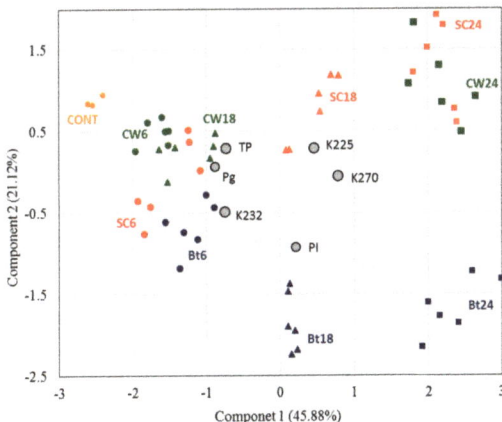

Fig. 3 Principal component analysis of the analysed parameters of olive oil stord with different closures (Cont- control, olive oil at the beguining of the test)

A characteristic olive oil spectrum is represented in Figure 4. These spectrum are similar to those presented by other authors [36–38] working with olive oil chemical characterization.

Cork Science and its Applications (CSA2017)   Materials Research Forum LLC
Materials Research Proceedings 3 (2017) 46-59   doi: http://dx.doi.org/10.21741/9781945291418-7

The prominent bands in the spectrum are located in the region of 2922 cm^{-1} due to the symmetrical and asymmetrical –CH– stretch vibration of the hydrocarbon chains, CH_2 and CH_3 aliphatic groups due to the alkyl rest of triacylglycerides.

In the region of 1744 cm^{-1} due to the carbonyl groups in the triacylglycerides. This absorbance band is characteristic to the oils with high short carbohydrate chain and saturated fatty acids content [36].

The spectral band at 1458 cm^{-1} is assigned to the vibrations of deformation C–H, related to total unsaturation assessing. However, this region is not usually used to discriminate different olive oils or to identify some olive oil adulterations [38].

The small band at 1370 cm^{-1} was assigned of methylene group (–CH_2–) deformation vibration and the band at 1234 cm^{-1} is assigned by the deformation vibration in the plan of the group =CH– [38][37].

The peak at 1157, 1029 and 721 cm^{-1} are assigned by the C=O bonds' vibration and to the C–C bonds vibration of carbohydrate chain from oil.

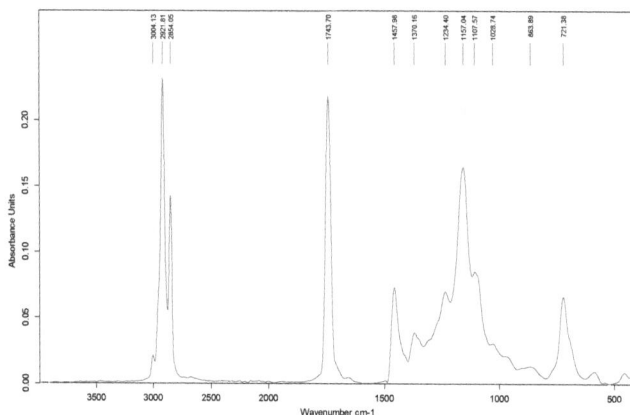

Fig. 4 Average FTIR-ATR spectrum of olive oil samples with the identification of the wavenumber of the more important peaks.

Several combinations of spectral region and pre-process were tested. For the Principal component analyses of the spectral information, the best results were obtained with a 2ndDer spectra and with a smooth of 25 as suggest by Santos et al. [39] for this pre-process (Figure 5). The spectral region selected for this analysis was 3049-2800 + 1777-1711+1309-1044 + 763-673 cm^{-1}. These regions were also used by Poiana et al [37] that worked with FTIR-ATR to differentiate among pure and adulterated oils and to study the thermooxidative processes in oils undergoing thermal stress. Other authors also use the mentioned region for the characterization of extra virgin olive oils from three cultivars in different maturation stages [40] and to classify and predict commercial grade of virgin olive oils [36].

The results expressed in Figure 5 are similar to those reported in Figure 3 but with some small differences. In fact in the PCA of Figure 3, only the chemical parameters related to the oxidative process were analyzed. In the values reported in Figure 3 we can conclude that some other important compounds maybe analyzed namely aldehydes, esters, alcohols and ketones

Cork Science and its Applications (CSA2017)   Materials Research Forum LLC
Materials Research Proceedings **3** (2017) 46-59   doi: http://dx.doi.org/10.21741/9781945291418-7

responsible for the most important flavors in olive oil [41]. However, those compounds have small changes which are confirmed by the similarity of the processed spectra represented in Figure 6.

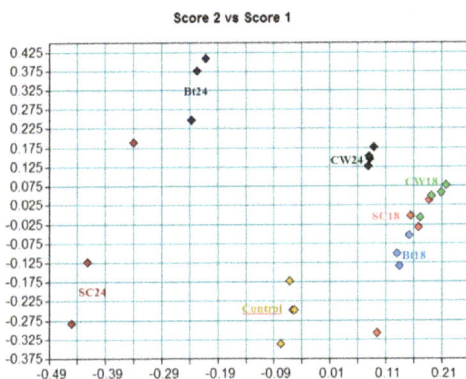

Fig. 5 Principal component analysis score plots of olive oils samples.

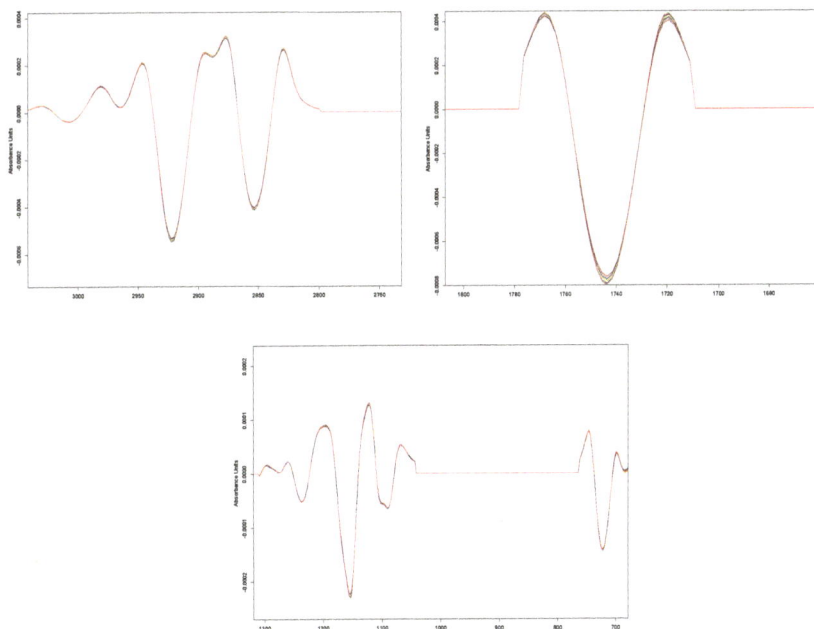

Fig. 6 Second derivative processed spectra from FTIR-ATR for all samples

Conclusion

Given the results shown in this study for the oxidative parameters, it was possible to conclude that the cork stopper has similar results to the screw cap during the storage time tested.

The olive oil sealed with a cork stopper with bee's wax for 18 months have a composition very similar to those present for other closures tested and more similar to the olive oil composition at the beginning of the test. The factor that has higher values of variance is the storage time, which means that the performance of the closures is quite similar. However, for Peroxide index and K270 the interaction between closure system and months is very significant.

Regarding the FTIR-ATR results is was possible to conclude that some small differences could be found in the different closure system, which may be related to the main volatile compounds responsible for olive oil flavor. This point will be the focus of future research.

Acknowledgement

Centro de Estudos Florestais is a Research Unit funded by FCT within UID/AGR/UI00239/2013.

LEAF, Linking Landscape, Environment, Agriculture and Food (UID/AGR/04129/2013).

Amorim Top Series Unit by the stopper availability.

This study was supported by Project CENTRO-04-3928-FEDER-000009 - Beira Baixa Terras de Excelência - Comunicação, Animação e Inovação.

The authors would like to express their gratitude to Isabele Salavessa (IPCB Languages centre) for the English revision.

References

[1] O. Anjos, C. Rodrigues, J. Morais, H. Pereira, Effect of density on the compression behaviour of cork, Mater. Des. 53 (2014) 1089–1096. https://doi.org/10.1016/j.matdes.2013.07.038

[2] O. Anjos, H. Pereira, M.E. Rosa, Effect of quality, porosity and density on the compression properties of cork, Holz Als Roh - Und Werkst. 66 (2008) 295–301. https://doi.org/10.1007/s00107-008-0248-2

[3] O. Anjos, H. Pereira, M.E. Rosa, Tensile properties of cork in the tangential direction: Variation with quality, porosity, density and radial position in the cork plank, Mater. Des. 31 (2010) 2085–2090. https://doi.org/10.1016/j.matdes.2009.10.048

[4] O. Anjos, H. Pereira, M.E. Rosa, Tensile properties of cork in axial stress and influence of porosity, density, quality and radial position in the plank, Eur. J. Wood Wood Prod. 69 (2011) 85–91. https://doi.org/10.1007/s00107-009-0407-0

[5] O. Anjos, H. Pereira, M.E. Rosa, Characterization of radial bending properties of cork, Eur. J. Wood Wood Prod. 69 (2011) 557–563. https://doi.org/10.1007/s00107-010-0516-9

[6] Á. García, O. Anjos, C. Iglesias, H. Pereira, J. Martínez, J. Taboada, Prediction of mechanical strength of cork under compression using machine learning techniques, Mater. Des. 82 (2015) 304–311. https://doi.org/10.1016/j.matdes.2015.03.038

[7] H. Pereira, Chemical composition and variability of cork from Quercus suber L., Wood Sci. Technol. 22 (1988) 211–218. https://doi.org/10.1007/BF00386015

[8] H. Pereira, Cork: biology, production and uses, Elsevier Science, 2007.

[9] S. Lequin, D. Chassagne, T. Karbowiak, J.-M. Simon, C. Paulin, J.-P. Bellat, Diffusion of Oxygen in Cork, J. Agric. Food Chem. 60 (2012) 3348–3356. https://doi.org/10.1021/jf204655c

[10] M.A. Fortes, M. Emilia Rosa, H. Pereira, The Cellular Structure of Cork from Quercus Suber L., IAWA J. 8 (1987) 213–218. https://doi.org/10.1163/22941932-90001048

[11] M.A. Silva, M. Julien, M. Jourdes, P.-L. Teissedre, Impact of closures on wine post-bottling development: a review, Eur. Food Res. Technol. 233 (2011) 905–914. https://doi.org/10.1007/s00217-011-1603-9

[12] P. Lopes, C. Saucier, Y. Glories, Nondestructive Colorimetric Method To Determine the Oxygen Diffusion Rate through Closures Used in Winemaking, J. Agric. Food Chem. 53 (2005) 6967–6973. https://doi.org/10.1021/jf0404849

[13] N. Kontoudakis, P. Biosca, R. Canals, F. Fort, J.M. Canals, F. Zamora, Impact of stopper type on oxygen ingress during wine bottling when using an inert gas cover, Aust. J. Grape Wine Res. 14 (2008) 116–122. https://doi.org/10.1111/j.1755-0238.2008.00013.x

[14] S. Caillé, A. Samson, J. Wirth, J.B. Diéval, S. Vidal, V. Cheynier, Sensory characteristics changes of red Grenache wines submitted to different oxygen exposures pre and post bottling, Anal. Chim. Acta. 660 (2010) 35–42. https://doi.org/10.1016/j.aca.2009.11.049

[15] V. Oliveira, P. Lopes, M. Cabral, H. Pereira, Influence of cork defects in the oxygen ingress through wine stoppers: Insights with X-ray tomography, J. Food Eng. 165 (2015) 66–73. https://doi.org/10.1016/j.jfoodeng.2015.05.019

[16] V. Oliveira, P. Lopes, M. Cabral, H. Pereira, Kinetics of Oxygen Ingress into Wine Bottles Closed with Natural Cork Stoppers of Different Qualities, Am. J. Enol. Vitic. 64 (2013) 395–399. https://doi.org/10.5344/ajev.2013.13009

[17] D. Boskou, Olive Oil Chemistry and Technology, Ilinois A, 2006. https://doi.org/10.1201/9781439832028

[18] A. Kiritsakis, A. Kanavouras, K. Kiritsakis, Chemical analysis, quality control and packaging issues of olive oil, Eur. J. Lipid Sci. Technol. 104 (2002) 628–638. https://doi.org/10.1002/1438-9312(200210)104:9/10%3C628::AID-EJLT628%3E3.0.CO;2-1

[19] R.J. Hamilton, The Chemistry of Rancidity in Foods, by John C. Allen (Editor), R.J. Hamilton (Editor), Aspen Publication, 1994.

[20] M. Servili, A. Taticchi, S. Esposto, B. Sordini, S. Urbani, Technological Aspects of Olive Oil Production, in: Olive Germplasm - Olive Cultiv. Table Olive Olive Oil Ind. Italy, InTech, 2012. https://doi.org/10.5772/52141

[21] F. Peres, L.L. Martins, S. Ferreira-Dias, Influence of enzymes and technology on virgin olive oil composition, Crit. Rev. Food Sci. Nutr. 57 (2017) 3104–3126. https://doi.org/10.1080/10408398.2015.1092107

[22] F. Angerosa, Influence of volatile compounds on virgin olive oil quality evaluated by analytical approaches and sensor panels, Eur. J. Lipid Sci. Technol. 104 (2002) 639–660. https://doi.org/10.1002/1438-9312(200210)104:9/10%3C639::AID-EJLT639%3E3.0.CO;2-U

[23] R. Aparicio, S.M. Rocha, I. Delgadillo, M.T. Morales, Detection of Rancid Defect in Virgin Olive Oil by the Electronic Nose, J. Agric. Food Chem. 48 (2000) 853–860. https://doi.org/10.1021/jf9814087

[24] Morales M. T.; Rios J.J.; Aparicio R., Changes in the Volatile Composition of Virgin Olive Oil during Oxidation: Flavors and Off-Flavors, J. Agric. Food Chem. 45 (1997) 2666–2673. https://doi.org/10.1021/jf960585+

[25] A. Kanavoouras, F. Coutelieris, Shelf-life predictions for packaged olive oil based on simulations, Food Chem. 96 (2006) 48–55. https://doi.org/10.1016/j.foodchem.2005.01.055

[26] A.I. Méndez, E. Falqué, Effect of storage time and container type on the quality of extra-virgin olive oil, Food Control. 18 (2007) 521–529. https://doi.org/10.1016/j.foodcont.2005.12.012

[27] G. Pristouri, A. Badeka, M.G. Kontominas, Effect of packaging material headspace, oxygen and light transmission, temperature and storage time on quality characteristics of extra virgin olive oil, Food Control. 21 (2010) 412–418. https://doi.org/10.1016/j.foodcont.2009.06.019

[28] European Union, Commission Implementing Regulation (EU) No 1348/2013 amending Regulation (EEC) No 2568/91, Off. J. Eur. Union. 2013 (2013) 31–67.

[29] M.L. Pizarro, M. Becerra, A. Sayago, M. Beltrán, R. Beltrán, Comparison of Different Extraction Methods to Determine Phenolic Compounds in Virgin Olive Oil, Food Anal. Methods. 6 (2013) 123–132. https://doi.org/10.1007/s12161-012-9420-8

[30] F. Peres, L.L. Martins, M. Mourato, C. Vitorino, S. Ferreira-Dias, Bioactive Compounds of Portuguese Virgin Olive Oils Discriminate Cultivar and Ripening Stage, J. Am. Oil Chem. Soc. 93 (2016) 1137–1147. https://doi.org/10.1007/s11746-016-2848-z

[31] J. Pokorny, L. Kalinova, P. Dysseler, Determination of chlorophyll pigments in crude vegetable oils. Results of a collaborative study and the standardized method, Pure Appl. Chem. 67 (1995) 1781–1787. https://doi.org/10.1351/pac199567101781

[32] F. Gutiérrez Rosales, S. Perdiguero, R. Gutiérrez, J.M. Olias, Evaluation of the bitter taste in virgin olive oil, J. Am. Oil Chem. Soc. 69 (1992) 394–395. https://doi.org/10.1007/BF02636076

[33] M.C. Peres, M. F., Henriques, L. R., Simões-Lopes, P., Pinheiro-Alves, Azeites da Galega Vulgar – Efeito do loteamento e do armazenamento, in: Actas Port. Hortic., 2009, (13), 186-191.

[34] F. Caponio, M.T. Bilancia, A. Pasqualone, E. Sikorska, T. Gomes, Influence of the exposure to light on extra virgin olive oil quality during storage, Eur. Food Res. Technol. 221 (2005) 92–98. https://doi.org/10.1007/s00217-004-1126-8

[35] M. Servili, R. Selvaggini, S. Esposto, A. Taticchi, G. Montedoro, G. Morozzi, Health and sensory properties of virgin olive oil hydrophilic phenols: agronomic and technological aspects of production that affect their occurrence in the oil., J. Chromatogr. A. 1054 (2004) 113–27. https://doi.org/10.1016/S0021-9673(04)01423-2

[36] A. Hirri, M. Bassbasi, S. Platikanov, R. Tauler, A. Oussama, FTIR Spectroscopy and PLS-DA Classification and Prediction of Four Commercial Grade Virgin Olive Oils from Morocco, Food Anal. Methods. 9 (2016) 974–981. https://doi.org/10.1007/s12161-015-0255-y

[37] M.-A. Poiana, E. Alexa, M.-F. Munteanu, R. Gligor, D. Moigradean, C. Mateescu, Use of ATR-FTIR spectroscopy to detect the changes in extra virgin olive oil by adulteration with soybean oil and high temperature heat treatment, Open Chem. 13 (2015).

[38] N. Vlachos, Y. Skopelitis, M. Psaroudaki, V. Konstantinidou, A. Chatzilazarou, E. Tegou, Applications of Fourier transform-infrared spectroscopy to edible oils, Anal. Chim. Acta. 573–574 (2006) 459–465. https://doi.org/10.1016/j.aca.2006.05.034

[39] O. Santos, A.J.A.; Caldeira, I; Anjos, Modelos de calibração de metanol e teor alcoólico em aguardentes por FTIR-ATR, in: Fórum ALABE 2016, 1–6.

[40] I. Gouvinhas, J.M.M.M. de Almeida, T. Carvalho, N. Machado, A.I.R.N.A. Barros, Discrimination and characterisation of extra virgin olive oils from three cultivars in different

Cork Science and its Applications (CSA2017)   Materials Research Forum LLC
Materials Research Proceedings **3** (2017) 46-59   doi: http://dx.doi.org/10.21741/9781945291418-7

maturation stages using Fourier transform infrared spectroscopy in tandem with chemometrics, Food Chem. 174 (2015) 226–232. https://doi.org/10.1016/j.foodchem.2014.11.037

[41] F. Peres, H.H. Jele, M.M. Majcher, M. Arraias, L.L. Martins, S. Ferreira-Dias, Characterization of aroma compounds in Portuguese extra virgin olive oils from Galega Vulgar and Cobrançosa cultivars using GC-O and GCxGC-ToFMS, Food Res. Int. 54 (2013) 1979–1986. https://doi.org/10.1016/j.foodres.2013.06.015

Cork Science and its Applications (CSA2017)　　　　　　　　Materials Research Forum LLC
Materials Research Proceedings 3 (2017) 60-74　　　　doi: http://dx.doi.org/10.21741/9781945291418-8

Using Cork Waste in Domestic Heating Equipment

J. I. Arranz[1, a *], F. J. Sepúlveda[1], M. T. Miranda[1], I. Montero[1], C. V. Rojas[1],
M. J. Trinidad[2]

[1]University of Extremadura. School of Industrial Engineering. Av. de Elvas, s/n, 06006 Badajoz, Spain

[2]Institute of Cork, Wood and Charcoal. C/Pamplona, s/n, 06800 Mérida, Spain

* jiarranz@unex.es

Keywords: Pellets, Combustion, Granulometric Separation Powder, Cork

Abstract. Biomass-fueled domestic heating equipment is increasingly being used, due to the low price of biofuels and the ease of finding them in distribution and sale networks. However, it is not possible to use just any type of biomass in these equipment, since its quality must be very high so that its performance is not too low. Also, the pelletization of the waste to be used is necessary, since a constant granulometry is necessary for the automatic feeding of the equipment.

Among the waste generated by the industries of granulated cork are grinding powder and granulometric separation powder. Of these two, only granulometric separation powder could be used as raw material for pellet manufacture for use in small domestic biomass equipment, due to its low ash percentage.

In this work, the combustion in a small domestic pellet stove is studied for different combinations of pellets made from granulometric separation powder and its behavior in this type of equipment.

For that purpose, five different types of pellets made from different granulometry combinations were used. A commercial Edilkamin Junior pellet stove with a power of 5.8 kW. Combustion gases, such as O_2, CO_2, CO, H_2, NO_x and SO_2, as well as combustion efficiency and chamber temperature were analyzed.

As most relevant results, it can be indicated that the stove combustion process did not present any anomaly of operation, during the realization of the tests. The maximum CO emissions barely reached the limits allowed and SO_2 emissions were practically zero. However, NO_x emissions were higher, and should be taken into account. On the other hand, the temperature values in the combustion chamber were significantly lower than those that can be reached with commercial pellets. Finally, efficiency values were very acceptable, with percentages above 96%.

Introduction

Biomass domestic heating equipment is increasingly being used, due to the low price of biofuels and the ease of finding them in the distribution and sale networks. However, it is not possible to use any type of biomass in these equipment, since its quality must be very high so that its performance is not low.

On the other hand, the most important residues in the industries of granulated cork are grinding powder and granulometric separation powder. Traditionally the term "cork powder" has been used to refer to all solid wastes from this type of industry [1]. However, each residue has specific characteristics depending on the stage of the production process where it takes place. Thus, grinding powder is generated in the stone mill, in order to uniform and regulate the size of the pieces of cork.

After the removal of grinding powder, a second selection of granulate, in this case by density, is carried out between the grains of pure cork and those bearing impurities. By using densimetric tables, combining vibration and air currents, the less dense grains (granulated cork) are separated from the denser ones, which constitute the residue "granulometric separation powder", which lacks the necessary qualities to be employed in granulated cork industries.

Both grinding powder and granulometric separation powder are susceptible to direct use as fuel. In fact, some granulated cork industries use both residues for the production of thermal energy.

Regarding granulometric separation powder, there are different sizes of rejection, depending on the product to be obtained. Thus, the most common granulated cork are those corresponding to 0.2-0.5 mm, 0.5-1 mm, 1-2 mm and 3-6 mm, whose nomenclature refers to the size limit values of particle.

In order to be able to use these by-products in domestic thermal generation equipment, their pelletization is necessary, since a constant particle size is required for the automatic feeding of the equipment and, when transporting them, it is essential to do it in the most efficient way. Thus, the densification of these products optimizes their logistics.

In addition, small domestic equipment can only use biofuels in the form of pellets. Although some "poly-fuel" equipments are currently on the market, they really are not, since they limit the entry of fuel into pellets and products with very uniform granulometry, such as olive stone.

Granulometric separation powder is one of the few residues that could be used as a pellet feedstock for use in small domestic biomass equipment because of its low ash content.

Several authors have carried out investigations of densification of cork. For example, Montero et al. [2] and Sepúlveda [3] manufactured pellets from different wastes of cork industry, stressing their characteristics and their corresponding densification, with the aim of making a specific use feasible. Nunes et al. [4] pelletized and characterized a product made from industrial cork waste, demonstrating the feasibility of obtaining a pellet with good physical and energy characteristics.

It is important to study the behavior of these by-products in small combustion equipment to verify if they can be used regularly.

Such studies are poorly developed. In this sense, different studies have shown that boiler characteristics, fuel properties and operating conditions significantly affect the combustion process and the amounts of polluting emissions [5] [6].

Also, pellets produced from different residues and under different preparation conditions have different chemical and physical characteristics, as well as a differentiated behavior during combustion under identical operating conditions. These characteristics significantly affect the emission of pollutants, even in those cases where the effect on the thermal efficiency of the boiler is small [7].

In this work, the combustion in pellet stove of different combinations of pellets made with granulometric separation powder is studied.

Cork Science and its Applications (CSA2017)   Materials Research Forum LLC
Materials Research Proceedings **3** (2017) 60-74   doi: http://dx.doi.org/10.21741/9781945291418-8

Materials and methods

Composition of pellets

Pellets of 6 mm diameter were manufactured in the Laboratory of Area of Machines and Thermal Motors at the University of Extremadura, using granulometric separation powder as raw material. Three types of granulometry were used: 0.5-1 mm, 1-2 mm and 2-3 mm, which are shown in Fig. 1.

Granulometric separation powder
0.5-1 mm

Granulometric separation powder
1-2 mm

Granulometric separation powder 2-3 mm

Fig. 1 Granulometric separation powder of different particle sizes.

The main characteristics of the pellets are shown in Table 1.

Cork Science and its Applications (CSA2017) Materials Research Forum LLC
Materials Research Proceedings **3** (2017) 60-74 doi: http://dx.doi.org/10.21741/9781945291418-8

Table 1. Properties of pellets from granulometric separation powder.

	Pellets 0.5-1 mm	Pellets 1-2 mm	Pellets 2-3 mm
Moisture (% wb)	8.02	7.96	6.53
Bulk density (kg/m^3 db)	697.02	740.32	698.54
Ultimate analysis			
C (% db)	50.50	52.97	54.52
H (% db)	5.80	6.15	6.61
O (% db)	38.43	36.35	33.38
N (% db)	0.43	0.38	0.75
S (% db)	0.03	0.02	0.00
Proximate analysis			
Volatile matter (% db)(	78.78	78.71	75.15
Fixed carbon (% db)	19.41	17.16	20.11
Ash (% db)	4.81	4.13	4.74
HHV (MJ/kg db)	21.41	24.21	21.56
Mechanical durability (%)	96.79	97.17	97.67
Fines (%)	0.01	0.02	0.01

Five combinations of different proportions of three types of granulometric separation powder (0.5-1, 1-2 and 2-3 mm) were carried out. In this way, the blend of products with similar energetic characteristics, but of different granulometry, will allow to study its effect on the densification of residues.

Table 2 shows the composition of the wastes used in each of the five combinations.

Table 2. Pellets composition.

	GSP1	GSP2	GSP3	GSP4	GSP5
Granulometric separation powder 0.5-1	50%	60%	40%	33%	50%
Granulometric separation powder 1-2 mm	50%	40%	60%	33%	0%
Granulometric separation powder 2-3 mm	0%	0%	0%	33%	50%

Equipment

In order to study the combustion of the manufatured pellets, a small commercial pellet stove Edilkamin Junior was used. Main characteristics of the stove are shown in Table 2.

Table 2. Characteristics of pellets stove.

Hopper capacity (kg)	15
Performance (%)	> 90
Output power min/max (kW)	3.4/5.8
Autonomy min/max (h)	10/22
Fuel consumption min/max (kg/h)	0.5/1.45
Flue gas duct diameter (mm)	80
Air inlet diameter (mm)	10

On the other hand, in order to determine the composition of the combustion gases, a TESTO 350 XL gas analyzer was used, which measures the amount of O_2, CO_2, CO, H_2, NO_x (corresponding to the sum of NO and NO_2 emissions) and SO_2.

A NICR / NI temperature probe, suitable for recording high temperatures, was used to obtain the temperature inside the stove.

In order to compare emissions, it was necessary to establish a specific O_2 reference level. In this work, a value of 11% was considered as reference, based on the following expressions:

$$CO[\text{ppm}\,11\%O_2] = \frac{CO\,[\text{ppm}]\cdot(21\text{-}11)}{(21\text{-}O_2[\%])} \tag{1}$$

$$H_2[\text{ppm}\,11\%O_2] = \frac{H_2\,[\text{ppm}]\cdot(21\text{-}11)}{(21\text{-}O_2[\%])} \tag{2}$$

$$NO_x[\text{ppm}\,11\%O_2] = \frac{NO_x\,[\text{ppm}]\cdot(21\text{-}11)}{(21\text{-}O_2[\%])} \tag{3}$$

$$SO_2[\text{ppm}\,11\%O_2] = \frac{SO_2\,[\text{ppm}]\cdot(21\text{-}11)}{(21\text{-}O_2[\%])} \tag{4}$$

Development of combustion tests

In order to carry out the combustion analysis, a five hour experiment was initially designed to ensure steady state. Subsequently, and given the acceptable stability reflected in the results, the duration of tests was reduced to three hours.

Pellets were manually loaded into the hopper, with approximate amounts of 10 kg for each case.

Likewise, the operating conditions of the stove were kept constant during all the tests. The thermal power selected was the minimum available, corresponding to 3.40 kW.

The pellets were heated by the introduction of hot air, while the fan provided primary air to the hearth of the stove. At an established temperature, the pellets began to release the first gases (light volatiles). Then, the first flame appeared, stabilizing the process after a short time. Once the steady state was reached, the fan was started to supply clean and hot air to the area to be heated.

The process was recorded by the probes described above. In all tests and despite the fact that the data were taken from the beginning of the tests, the data corresponding to the first 30 minutes were not taken into account, where important variations occur that influence the estimation of the average values of the emissions. In this way, both tables and graphs with the results refer to the process of combustion of the products after their stabilization.

Combustion efficiency

The fundamental objective of a combustion process is to release the calorific value contained in the fuel used, obtaining mainly CO_2, H_2O and other by-products. In order to carry out a quantitative comparison of the combustion quality, the combustion efficiency was used as the reference index.

A combustion process can be considered ideal when no incomplete combustion products are obtained. This means that, in the particular case of carbon, the entire content of the carbon must be transformed into CO_2. Based on this definition, the combustion efficiency (τ) can be evaluated according to the expression Eq. 5.

Cork Science and its Applications (CSA2017)   Materials Research Forum LLC
Materials Research Proceedings **3** (2017) 60-74   doi: http://dx.doi.org/10.21741/9781945291418-8

$$\tau = \frac{CO_2 \; real}{CO_2 \; theoretical \; maximum} \cdot 100 \tag{5}$$

Other authors, such as Sevon and Cooper [8], define the efficiency as a function of CO and CO_2 concentrations in the exhaust gases, using the expression Eq. 6.

$$\tau = \frac{CO_2[\%]}{CO_2[\%] + \left(\dfrac{CO[ppm]}{10.000}\right)} \cdot 100 \tag{6}$$

In the present work the expression Eq. 6 was chosen to assess the efficiency of the process.

Results
Analysis of combustion of pellets
To carry out the tests the pellets were selected, whose ash contents (slightly higher than 3% db) guaranteed the proper functioning of the equipment. They were also pellets composed of granulometric separation powder, one of the two major residues that would enable densification on a large scale.

In the following sections, the analysis performed, as well as the graphs of variation of the most significant parameters, are shown.

It is noted that the results are very similar for all tests. This is why in this paper we will analyze only the first two (GSP1 and GSP2), showing a table with a summary of all the results. Combustion of GSP1 pellets.

Fig. 2 shows the percentages of O_2, CO_2, CO, H_2, NO_x and SO_2 recorded in the flue gases of GSP1 pellets. During the test, high O_2 values were recorded due to the excess air that the stove introduces into the combustion chamber. During the initial period there was a rapid decrease in O_2 concentration, as a consequence of a greater inlet of pellets from the screw that communicates the hopper and the chamber [5]. This decrease was also observed by Hays et al [9] during simulations of herbaceous waste burning.

Usually, in domestic stoves, high excess of air is obtained, since they are equipment designed for small rooms, which do not require high temperatures in the air used to heat them, while avoiding the fumes being expelled by the chimney at excessive temperatures.

The concentration of CO_2 underwent an opposite process, that is, a slight rise in the initial phase of the test. Then, both parameters were stabilized, reaching average values of 18.5 and 1.8% for O_2 and CO_2, respectively. Thus, the achievement of such high values of O_2 is linked to obtaining very low values of CO_2 during combustion [10].

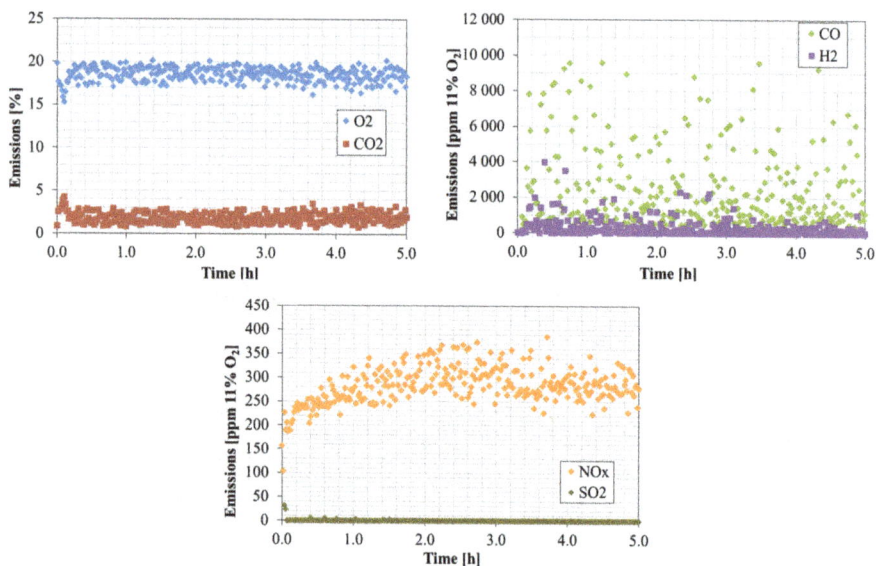

Fig. 2 O_2, CO_2, CO, H_2, NO_x and SO_2 emissions during combustion of GSP1 pellets.

High dispersion in CO concentration was observed, while H_2 emissions were more or less constant with some isolated peaks.

The emissions of nitrogen oxides increased slightly during the five hours of the test, with a greater dispersion in the central part of the test. The mean value recorded was 290 ppm (11% O_2).

Finally, SO_2 concentration remained practically null, with some values isolated at the beginning of the test. The presence of specific emissions of SO_2 during the first instants, that is, during the stage of greater instability of the tests, was a trend that was repeated for all the products.

Combustion of GSP2 pellets
Fig. 3 shows the percentages of O_2, CO_2, CO, H_2, NO_x and SO_2 recorded in the combustion gases of the GSP2 pellets. As in the previous test, there was a rapid decline in O_2 concentration and an increase in the concentration of CO_2 at the start of combustion, then stabilized for average values of 18.7 and 1.7% for O_2 and CO_2, respectively.

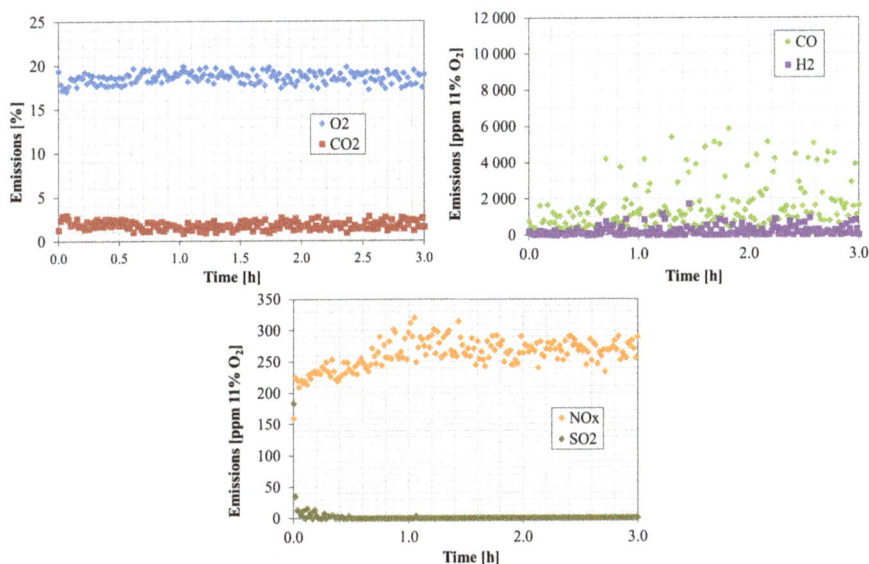

Fig. 3 O_2, CO_2, CO, H_2, NO_x and SO_2 emissions during combustion of GSP2 pellets.

High dispersion is also noted in the case H_2 and, especially, CO. The mean values recorded were 1 821 and 233 ppm (11% O_2), respectively.

Nitrogen oxides increased slightly during the first hour of the test, from values close to 230 ppm (11% O_2) to 300 ppm (11% O_2), stabilizing for an average value of 268 ppm (11% O_2). On the other hand, practically no emissions of SO_2 were detected, except in the first half hour of the test (insignificant values).

Summary of results and considerations
Table 3 shows the mean values obtained during the tests.

Table 3. Results of combustion of pellets.

Emissions	GSP1	GSP2	GSP3	GSP4	GSP5
O_2 [%]	18.5	18.7	19.5	19.8	19.9
CO_2 [%]	1.8	1.7	1.2	1.0	1.1
CO [ppm 11% O_2]	2 504.6	1 820.7	1 633.7	2 163.2	1 778.8
H_2 [ppm 11% O_2]	275.4	233.4	249.3	277.9	283.3
NO_x [ppm 11% O_2]	289.8	268.2	248.8	285.4	288.1
SO_2 [ppm 11% O_2]	0.1	0.0	0.8	0.0	0.8

In general, it was observed that when more excess air is introduced (greater percentage of O_2), there is more air volume and, therefore, the percentage value of CO_2 decreases, as shown in Fig. 7.

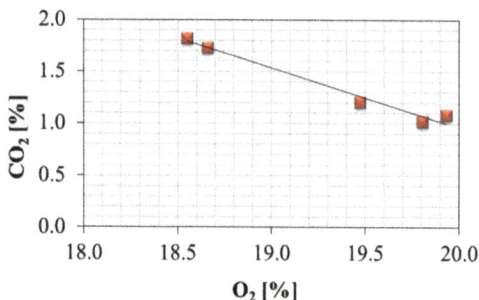

Fig. 7 Relationship between O_2 and CO_2 emissions.

O_2 concentration in the flue gas provides a measure of the excess air introduced into the stove. The combustion of the GSP5 pellets was carried out with a higher air excess (19.9%), followed by the combustion of pellets GSP4 and GSP3, with average values very close to each other. Likewise, during the combustion of GSP1 and GSP2 pellets the lowest results were obtained (18.5 and 18.7%, respectively).

Fig. 8 shows graphically CO, H_2 and NO_x emissions recorded during the combustion tests. SO_2 emissions were not represented due to their low amount.

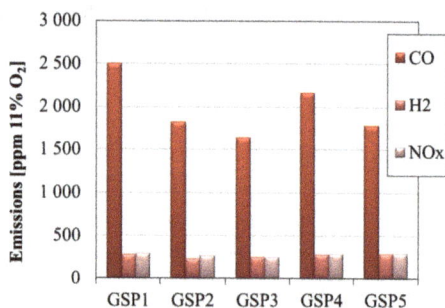

Fig 8 CO, H_2 and NO_x emissions.

O_2 emissions from a combustion do not only affect the proportions of CO_2, but also those of CO and NO_x. Arranz [10] observed an exponential increase in CO emissions as the proportion of O_2 increased during the combustion of cork pellets, as well as a linear increase in NO_x emissions.

Similar mean CO values were obtained in GSP2, GSP3 and GSP5 combustion tests and higher for the GSP1 and GSP4 experiments.

UNE-303-5 [11] limits CO emissions to 3 000 mg/m^3 (10% O_2) in class 3 boilers with a power of less than 50 kW, equivalent to 2 383 ppm (11% O_2). This limit was slightly exceeded by GSP1 pellets. In all other cases, CO emissions were below the maximum allowed. However, it is necessary to indicate that the combustion tests were carried out at the minimum power of the boiler (3.4 kW), and lower mean concentrations could be obtained in tests closer to the nominal power of the equipment, that is, the power at which it was dimensioned [10]. In addition, this type of small power equipment does not allow to control the excess of air introduced and work (independently of the load) with quite high values. Thus, Rabaçal et al [12] conclude that emissions caused by incomplete combustion, especially CO emissions, can be minimized by optimizing the boiler operating conditions, in particular by controlling excess air.

Deficient combustion results in high CO emissions and may be due to different causes, such as low combustion temperature, insufficient oxygen, poor fuel mixture with air or too short residence time of the combustion gases in the chamber, among others [13] [14].

Comparing the results obtained with those published by other authors, Mediavilla et al [15] carried out combustion tests with pellets from combinations of vine and cork residues in a 17.5 kW boiler. The tests were designed for a duration of 10 hours, although they were reduced due to the accumulation of ashes and to the formation of slag in the burner. Combustion of vine shoots pellets produced the highest CO emission values (6 545 ppm 11% O_2), followed by the combustion of cork waste pellets (2 254 ppm 11% O_2) and pine sawdust pellets (349 ppm 11% O_2). Pellets from combination of shoots and cork residues showed lower CO emissions to a higher percentage of their composition, reaching a minimum value of 436 ppm (11% O_2) for a mixture with 70% cork residues.

On the other hand, Fernandes et al [16] and Verma et al [13] recorded CO emissions of 1 438, 1 718 and 1 180 ppm (11% O_2) in the combustion of commercial pine pellets, straw and peat waste, respectively, in boilers of 10 and 40 kW. In both cases, the use of boilers of greater power prevents a more exhaustive comparison of the combustion of the different residues.

It is also necessary to evaluate the process from the environmental point of view, based on possible generation of nitrogen oxides and the presence of sulfur oxides in the flue gases. In terms of NO_x emissions, Verma et al [17] established that, in domestic boilers, where the temperature of the combustion chamber hardly reaches values above 1 300 °C, NO generation is the main source of NO_x emissions. In turn, the emission of NO_x depends on the nitrogen content in the fuel and increases with the content of this and with the concentration of oxygen during the combustion [18] [19].

Returning to the work of Mediavilla et al [15], the highest NO value corresponded to the combustion of pellets of cork waste, followed by mixtures of cork waste and vine shoots, pellets of vine shoots and, finally, pine sawdust pellets. The proportion of nitrogen content and NO emissions was maintained for the densified independent residues (pine sawdust, cork residues and vine shoots), altering in the case of combinations. Combustion of cork waste and vine shoots pellets showed NO emissions of 229 and 157 ppm (11% O_2), reducing the value with increasing the proportion of waste cork in the mixture.

This situation was also observed by Miranda et al [20], who carried out pelletizing and combustion tests on blends of olive pomace and pyrenean oak residues. Thus, while pellets composed of 50% of each residue and with a nitrogen content of 1.07% (db) showed NO_x emissions of 327 ppm 11% O_2 on average, olive pomace pellets with nearly twice nitrogen content (1.98% db), emitted approximately 324 ppm (11% O_2).

Cork Science and its Applications (CSA2017)   Materials Research Forum LLC
Materials Research Proceedings 3 (2017) 60-74   doi: http://dx.doi.org/10.21741/9781945291418-8

In the present work, the characterization of the granulometric separation powder pellets showed different percentages of nitrogen in each case, with a maximum value of 0.75% (db) for the corresponding to 2-3 mm. However, in the combustion of the blends made with this residue no much higher amounts of NO_x were recorded compared to the other combinations, due in part to the fact that the differences in nitrogen content of the densified products were not significant.

Therefore, NO_x emissions were higher than those reported for pine sawdust pellets (44 ppm 11% O_2) and pellets of combinations of cork residues and vine shoots (137-146 ppm 11% O_2) and below than blends of pyrenean oak and olive pomace pellets (327 ppm O_2 11%).

In all cases, there was a certain tendency to increase NO_x emissions throughout the combustion process, a situation that were also noticed by Arranz [10] and Miranda [20]. The evolution of NO_x underwent a slight increase in all experiments. This tendency may be due to the increase of the temperature in the chamber and the increase of excess air when the combustion is stabilized.

Other authors, such as Fernandes et al [16], recorded very low emissions of NO_x in the combustion of commercial pine pellets, lower in all cases at 37 ppm (11% O_2); Verma et al [13] reported 88 and 46 ppm (11% O_2) for straw and peat residue pellets, respectively.

Sulfur oxides are due exclusively to the fuel composition. In the case of the residues analyzed, the sulfur content is negligible, so SO_2 appears in minimal amounts in the flue gases.

On the other hand, the control of the temperature in the combustion chamber is an important factor in the combustion of biomass products. Obtaining exceptionally high values can lead to the appearance of agglomerated ashes, which can clog the burner holes and prevent the entry of air into the chamber, with consequent loss of performance of the equipment.

Fig. 9 shows the distribution of the temperatures recorded in the combustion chamber throughout the different tests performed. In all cases, three zones can be seen: the first one, lasting approximately 20 minutes, in which the temperature rises rapidly to reach values between 250 and 300 °C; the second stage, more or less stable, in which the temperature undergoes slight variations and extends to the middle of the experiment (1.5 hours); and the third one, in which the temperature oscillates considerably and in which there was a greater dispersion of the data, possibly due to the accumulation of ashes and unburned fuel in the combustion chamber.

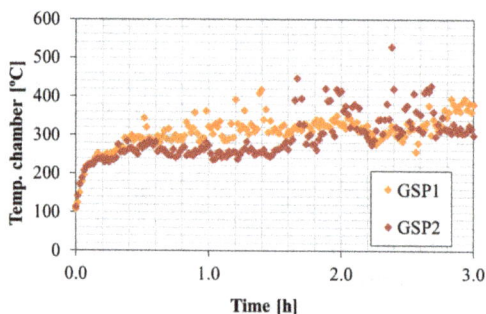

Fig. 9 Combustion chamber temperatures during the GSP1 and GSP2 tests.

Excluding the first half hour of the tests, mean and maximum temperatures recorded are shown in Table 4.

Table 4. Temperatures recorded during combustion tests [°C].

	GSP1	GSP2	GSP3	GSP4	GSP5
Mean value	322	303	326	314	273
Maximum value	417	529	533	510	444

The maximum temperature values refer to specific times of the test and usually correspond to a sudden fall of the pellets in the combustion chamber. Thus, in spite of a uniform distribution of densified products, the presence of pellets of greater or lesser length affects the transport of the fuel by the screw of the stove, causing slight variations in the feeding.

With regard to the average values of temperature, the maximum was recorded during the combustion of GSP3 pellets, product composed of granulometric separation power 0.5-1 mm and 1-2 mm. The combustion of lower density pellets means a lower availability of fuel for burning, so that at similar heating values, the temperatures recorded in the combustion chamber will necessarily be lower. However, given the small variation between these two magnitudes in the pellets manufactured, no substantial differences in temperature were observed.

Arranz [10] obtained approximate temperature values of 477 °C during the combustion of commercial pine wood pellets, in a stove of similar characteristics and at minimum power. These pellets, with a low heating value of 17.5 MJ/kg (wb) and bulk density 730 kg/m^3 (wb), reached higher temperatures in the combustion chamber, on the order of 1.5 times more than those recorded for pellets of cork waste. Since the heating value of the latter was between 17.6 and 19.3 MJ/kg (wb), lower chamber temperatures may be due to lower bulk density (693 kg/m^3 (wb) on average).

In terms of combustion process efficiencies, Table 5 shows the average percentages achieved during the execution of the different tests.

Table 5. Average efficiencies [%].

	GSP1	GSP2	GSP3	GSP4	GSP5
Average efficiency	96.5	97.2	97.9	97.2	96.0

In general, the determined efficiency values were high, with percentages higher than 96%, due to the small amount of gaseous unburned present in the combustion. Thus, a higher value of this parameter means that the combustion has been developed more completely. The maximum were recorded for the combustion of the GSP3 pellets, characterized by a lower amount of CO. Fig. 10 represents the relationship between both variables in performed tests.

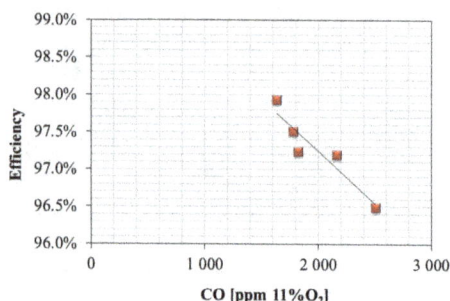

Fig. 10 Relationship between CO emissions and combustion process efficiency.

Conclusions

The main conclusions drawn from the combustion tests are as follows:

- The combustion of the pellets did not show any anomaly in relation to ash accumulation and overflow of the burner, being able to be carried out without interruption in tests of 3 and 5 hours of duration.
- The maximum CO emissions allowed by UNE-EN-303-5 [11], set at 2 383 ppm (11% O_2), were exceeded only during the combustion of the GSP1 pellets at minimum boiler power. Lower concentrations can be obtained in situations closer to the nominal power of the equipment.
- NO_x emissions were higher than those recorded for commercial pellets, with a slight tendency to increase emissions throughout the combustion process.
- SO_2 emissions were practically zero in all analyzed pellets.
- There were no substantial differences in temperature in the combustion chamber during the tests performed, possibly due to the similarity between bulk densities and heating values of the tested products. Likewise, the values recorded, significantly lower than those reached by commercial pellets, do not make possible the melting of the ashes or the appearance of other derived problems.
- The determined efficiency values were high, with percentages above 96% due to the small amount of gaseous unburned present in the combustion. The maximum was recorded for the combustion of GSP3 pellets, characterized by a lower amount of CO emissions.

In general, it can be concluded that pellets made from granulometric separation powder, in their different blends, can be used in a commercial pellet stove.

Since this type of small equipment is very limited in its efficiency performance, it could be said that the use of these pellets would be completely possible in boilers of greater power, with better characteristics in terms of performance and operational control.

References

[1] L. Gil, Cork powder waste: an overview, Biomass and Energy 13 (1997) Nos 1/2 59-61.

[2] I. Montero, M.T. Miranda, F.J. Sepúlveda, J.I. Arranz, M.J. Trinidad, C.V. Rojas, Analysis of pelletizing of wastes from cork industry, Dyna Energ. Sostenibilidad 3 (2014).

[3] F.J. Sepúlveda, Selective Use for the Integral Valorization of Wastes from Cork Industry. Ph.D. Thesis, University of Extremadura, Badajoz, Spain, 2014.

[4] L.J.R. Nunes, J.C.O. Matias, J.P.S. Catalão, Energy recovery from cork industrial waste: Production and characterisation of cork pellets, Fuel 113 (2013) 24-30. https://doi.org/10.1016/j.fuel.2013.05.052

[5] J. Dias, M. Costa, J.L.T. Azevedo, Test of a small domestic boiler using different pellets, Biomass Bioenergy 27 (2004) 531-539. https://doi.org/10.1016/j.biombioe.2003.07.002

[6] O. Sippula, K. Hytönen, J. Tissari, T. Raunemaa, J. Jokiniemi, Effect of wood fuel on the emissions from a top-feed pellet stove, Energy Fuel 21 (2007) 1151-1160. https://doi.org/10.1021/ef060286e

[7] J.F. González, C.M. González-García, A. Ramiro, J. Gañán, A. Ayuso, J. Turegano, Use of energy crops for domestic heating with a mural boiler, Fuel Processing Technology 87 (2006) 717–726. https://doi.org/10.1016/j.fuproc.2006.02.002

[8] D.W. Sevon, D. Cooper, Modelling combustion efficiency in a CFB liquid incinerator, Chemical Engineering Science 46 (12) (1991) 2983-2996. https://doi.org/10.1016/0009-2509(91)85003-G

[9] M. D. Hays, P. M. Fine, C D. Geron, M. J. Kleeman, B. K. Gullett, Open burning of agricultural biomass: Physical and chemical properties of particle-phase emissions, Atmospheric Environment 39 (2005) 6747-6764. https://doi.org/10.1016/j.atmosenv.2005.07.072

[10] J.I. Arranz, Análisis del densificado de la combinación de diferentes residuos biomásicos. Doctoral Thesis. E.I.I., Badajoz, 2011.

[11] AENOR. UNE-EN-303-5. Calderas de calefacción. Parte 5: Calderas especiales para combustibles sólidos, de carga manual y automática y potencia útil nominal hasta 500 kW, 2013.

[12] M. Rabaçal, U. Fernandes, M. Costa, Combustion and emission characteristics of a domestic boiler fired with pellets of pine, industrial wood wastes and peach stones, Renewable Energy 51 (2013) 220-226. https://doi.org/10.1016/j.renene.2012.09.020

[13] V. K. Verma, S. Bram, F. Delattin, P. Laha, I. Vandendael, A. Hubin, J. De Ruyck, Agro-pellets for domestic heating boilers: Standard laboratory and real life performance, Applied Energy 90 (2012) 17-23. https://doi.org/10.1016/j.apenergy.2010.12.079

[14] V.K. Verma, S. Bram, G. Gauthier, J. De Ruyck. Evaluation of the performance of a multi-fuel domestic boiler with respect to the existing European standard and quality labels: part-1, Biomass Bioenergy 35.1 (2011) 80–9. https://doi.org/10.1016/j.biombioe.2010.08.028

[15] I. Mediavilla, M.J. Fernández, L.S. Esteban, Optimization of pelletisation and combustion in a boiler of 17.5 kWth for vine shoots and industrial cork residue, Fuel Process. Technol. (2009) 90:621–8. https://doi.org/10.1016/j.fuproc.2008.12.009

[16] U. Fernandes, M. Costa. Particle emissions from a domestic pellets-fired boiler, 4th International Congress on Energy and Environment Engineering and Management. Mérida (Spain), 2011.

[17] V.K. Verma, S. Bram, G. Gauthier, J. de Ruyck. Performance of a domestic boiler as a function of operational loads: part-2. Biomass Bioenergy 35 (2011) 272-279. https://doi.org/10.1016/j.biombioe.2010.08.043

[18] O. Pastre, Analysis of the technical obstacles related to the production and utilization of fuel pellets made from agricultural residues, Pellets for Europe, Altener (2002).

Cork Science and its Applications (CSA2017)   Materials Research Forum LLC
Materials Research Proceedings 3 (2017) 60-74   doi: http://dx.doi.org/10.21741/9781945291418-8

[19] H. Wiinikka, R. Gebart, Critical parameters for particle emissions in small-scale fixed-bed combustion of wood pellets, Energy Fuel 18 (2004) 897-907. https://doi.org/10.1021/ef030173k

[20] T. Miranda, J.I. Arranz, I. Montero, S. Román, C.V. Rojas, S. Nogales, Characterization and combustion of olive pomace and forest residue pellets, Fuel Processing Technology 103 (2012) 91–96. https://doi.org/10.1016/j.fuproc.2011.10.016

Cork Science and its Applications (CSA2017)
Materials Research Proceedings 3 (2017) 75-83

Materials Research Forum LLC
doi: http://dx.doi.org/10.21741/9781945291418-9

The Use of Cork Waste as a Sorbent for Pesticides and Heavy Metals Generated During the Wine Manufacturing Process

P. Jové[1,a*], N. Fiol[2,b], I. Villaescusa[2,c], M. Verdum[1,d], L. Aguilar[3,e], C. Bosch[4,f], J. Morató[3,g]

[1]Catalan Cork Institute. Miquel Vincke i Meyer, 13 – 17200 Palafrugell, (Girona)

[2]University of Girona. Department of Chemistry Engineering. Campus Montilivi, (Girona)

[3]Unesco Chair on Sustainability, Polytechnic University of Catalonia, Terrassa, Spain

[4]Fundació CTM Centre Tecnològic, Unit Sustainability of Catalonia, Manresa, Barcelona, Spain

[a]pjove@icsuro.com, [b]nuria.fiol@udg.edu, [c]isabel.villaescusa@udg.edu, [d]mverdum@icsuro.com, [e]lorena.aguilar@upc.edu, [f]fcarme.bosch@ctm.com.es, [g]jordi.morato@upc.edu

Keywords: Cork Waste, Biosorbent, Pesticides, Heavy Metals, Water Treatment, Quercus Suber

Abstract. The aim of this study was to evaluate the adsorption capacity of cork wastes for the pollutants generated during wine manufacturing process. Adsorption was focused on four pesticides (aldrin, chlorpyrifos, metalaxyl and tebuconazole) and two heavy metals (Cu (II) and Ni (II)). The final purpose is to use this natural adsorbent as a substrate of a constructed wetland to improve its efficiency as wastewater treatment system. The high efficiency of cork as a sorbent of these pollutants is shown by the fact that equilibrium contact time obtained was 40 minutes. The highest adsorption capacities were exhibited for chlorpyrifos in the case of pesticides and for Ni (II) in the case of heavy metals. Experimental constructed wetlands filled with cork showed great removal efficiencies for these pesticides (more than 95%).

This study demonstrates that cork waste is a potential sorbent for some pesticides and heavy metals and may have relevance in the future treatment of pollutants-contaminated waters.

Introduction

Development of technically simple and economically attractive methods of industrial wastewaters purification is one of the most important priorities of 21st century (Volesky, 2001). Therefore, the best solution is preventing the entrance of pollutants in the ecosystem.

Conventional technologies traditionally used for the removal of some pollutants from aqueous solutions are expensive and becomes less effective. The use of sorbents for the removal of toxic pollutants is one of the most recent developments in environmental water treatments. The major advantages of this technology over conventional ones include its high efficiency and the minimization of chemical or biological sludge (Park et al., 2010).

On the other hand, constructed wetlands are designed to use the natural processes involving wetland vegetation, soils, and the associated microorganisms to assist in treating wastewaters. The combination of both systems, sorbents and constructed wetlands, aims to increase the efficiency of wastewater treatment. The use of natural sorbents such as cork waste also keeps the sustainability of the process.

Cork is the bark of the cork oak tree (*Quercus suber* L) and is natural, renewable and biodegradable raw material. It has a combination of properties that make it unique and versatile.

Cork Science and its Applications (CSA2017)   Materials Research Forum LLC
Materials Research Proceedings 3 (2017) 75-83   doi: http://dx.doi.org/10.21741/9781945291418-9

Cork oak forests extend over an area of almost 2.2 million hectares, concentrated mainly in the Mediterranean region, in the South of Europe and in North Africa (Portugal, Spain, Italy, France, Algeria, Morocco, and Tunisia). Europe has about 60 % of the total production area (cork forests) and produces more than 80 % of the world's cork. The main use of cork is the manufacture of cork stoppers but not all stripped cork is suitable for that activity. This material is called cork waste and could be used as a sorbent. Different types of biomasses (or sorbents of natural origin) have been studied for the last two decades and the sorption characteristics of many of them have been widely investigated. There are few studies on the ability of cork as a sorbent but these indicate that it might be a good material for its activity (Domingues, 2005, 2007, Chubar et al. 2003, 2004; Hanzlík et al. 2004; Villaescusa et al. 2000,2002).

Winery wastewaters are generated by various processes and operations carried out in wine production. This wastewater is characterized by the high content of organic matter, suspended soils and large variations in a seasonal flow production. Also, heavy metals like copper (Cu) and Nickel (Ni) and some pesticides such as chlorpyrifos and tebuconazole may be present in wastewater generated from the wineries as a result of the control of parasites of grape. These compounds can pass into the environment depending on the technological process (production and waste treatment) and the concentration factor of the fruit (Cabras and Angioni, 2000).

Constructed wetlands may offer an efficient low-cost, low-maintenance and energy alternative for wastewater treatment. Constructed wetlands also have the advantage of being able to accept seasonal flow fluctuations without adversely affecting the functional aspects of the treatment system so, is an alternative for wineries. The use of cork waste as a filling of a constructed wetland tries to be an improvement of the system.

The aim of this work was to evaluate the sorption ability of cork waste for some pollutants (pesticides and heavy metals) generated during wine manufacturing process using batch experiments and a pilot scale constructed wetland filled with cork waste.

Material and methods
Reagents
Heavy metals analytical standards were: Copper (II) chloride dihydrate extra pure (Sharlab) and Nickel (II) chloride hexahydrate (Sharlab). Pesticides analytical standards were purchased from Sigma-Aldrich: Aldrin, Chlorpyrifos, Tebuconazole, Metalaxyl and Phenanthrene-d10.To prepare the standard solutions acetone and methanol 215 SpS (Teknocroma) were used.

Cork samples
A cork factory supplied the cork waste. Three granulometric fractions: 2-3mm, 3-7mm and 15-20mm were used to select the optimal particle size for the adsorption process.

Batch sorption experiments
In the case of heavy metals, granulate cork (0.06g) is put in contact with 20 mL of solution. The mixture is set under stirring with Rotator Drive STR4 (Stuart Scientific). Once the set time elapsed, aqueous solution was filtered with a cellulose filter. pH was determined with a pH meter Crison BasiC 20. Concentration of metal was analyzed by flame atomic absorption (Varian 220 FS). Calibration standards suitable for measuring the concentration of heavy metals have been used. The appropriate hollow cathode lamp was also used.

In the case of pesticides, granulate cork (0.3g) is put in contact with 100 mL of solution. The solutions were mixed with a Vibromatic oscillating shaker at 700 oscillations/min. Concentration of pesticides were analyzed by solid phase microextraction (SPME) procedure and gas chromatography-mass spectrometry (GC-MS) analysis (GC-MS-SPME). For SPME extraction, 18 mL of each sample were analyzed. The fiber was immersed into the aqueous phase with

Cork Science and its Applications (CSA2017)   Materials Research Forum LLC
Materials Research Proceedings 3 (2017) 75-83   doi: http://dx.doi.org/10.21741/9781945291418-9

agitation at 50°C during 30 min. After extraction, the fiber was thermally desorbed for 10 min into the liner at 250°C. The splitless time was set at 4 min and desorption time at 10 min. GC was performed with 6890N Agilent chromatograph equipped with a MSP2 Gerstel autosampler and coupled to a MS 5973N mass spectrometer. The separation was achieved using an HP-5MS column (30m, 0.25 mm, 0.25 µm, film thickness) (J&W Scientific, Folsom, CA, USA), and the CG program was: 60°C (1min), increased by 25°C/min to 163°C, increased by 0.3°C/min to 167°C, increased by 30°C/min to 210°C (held for 2 min), increased by 5°C/min to 250°C (held for 4min). The carried was helium (99.999%) from Abello Linde with a constant flow rate of 2mL/min. The mass spectrometer was operated in a selected ion monitoring mode (SIM). The monitored ions were of m/z: aldrin 263 and 293; chlorpyrifos 197 and 314; tebuconazole 125 and 250; metalaxyl 250 and 125. The quantification of pesticides was based on comparison of the areas for the monitored molecular ions to that of the internal standard, with calibration response curves generated from six different concentrations characteristic of each pesticide (Mühlen C et al. 2014 and Scheyer A et al. 2005).

Batch experiments were conducted to study the effect of cork particle size, pH (in the case of heavy metals), contact time, amount of organic material (using COD value) and temperature in sorption process. The last two simulates the conditions that could occur during the treatment of water in a wetland. The effect of the presence of organic material or the chemical oxygen demand (COD) value in adsorption process was determined using a synthetic wastewater. This solution was prepared in accordance with the table 1.

Table 1. Composition of synthetic wastewater.

Reagents	Composition 1	Composition 2	Composition 3
Sugar (g/L)	0.3	0.3	0.3
Basic fertilizer (ml/L)	2	3	4
COD (mg/L)	510	574	598

In both cases, heavy metals and pesticides, the results are expressed in percentage of removal and coefficient of variation values.

Constructed wetland filled with cork waste
A first test on the ability of cork waste as a sorbent and a filling of a wetland was carried out. Three constructed wetlands, two horizontal (HW1 and HW2) and one vertical (VW) filled with cork waste were monitored during 16 days, at a mean hydraulic load of 40 ± 4 L/(m²/day) (VH) and 20 ± 4 L/(m²/day) (HW1 and HW2). The flown water was winery wastewater with a secondary treatment.

There were two tanks, a control tank with winery wastewater and a pollution tank with spiked winery wastewater: 0.5 mg/L of chlorpyrifos, tebuconazole, metalaxyl, Cu (II) and Ni (II). One horizontal (HW1) and vertical (VW) wetland worked with periods of contamination being fed with the pollution tank and decontamination periods being fed with the control tank. First 8 days, HW1 and VW were fed with the pollution tank and last 8 days were fed with the control tank. The other horizontal wetland (HW2) only operated with the control tank. The sampling was carried out in feeding tank and in the exit of the wetlands. The results are expressed in µg/L and removal percentages.

Cork Science and its Applications (CSA2017)
Materials Research Proceedings 3 (2017) 75-83

Materials Research Forum LLC
doi: http://dx.doi.org/10.21741/9781945291418-9

Concentration of pesticides was analyzed by the same methodology described above and concentration of metal was analyzed by an inductively coupled plasma mass spectrometry (ICP-MS).

Results and discussion
The first part of the study was determined the best optimal conditions for target pollutants adsorption such as cork particle size, contact time, pH (in the case of heavy metals), composition and temperature.

Effect of particle size
The effect of cork particle size on the adsorption of Cu (II) was investigated and the results are presented in Figure 1A. Medium particle size (3-7mm) and small particle size (2-3mm) presented similar results with percentage of removal higher than 50%. The same has been observed in the case of chlorpyrifos where the percentage of removal increased as smaller was the particle size (from 75.4 to 100%) (Fig. 1B). It was chosen the smallest particle size (2-3mm) to continue the study.

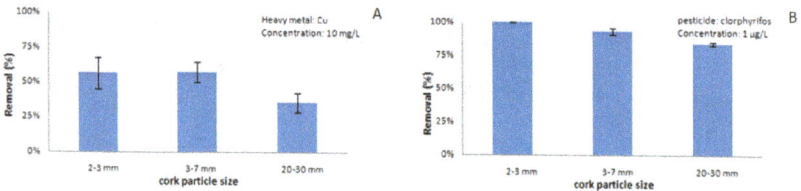

Fig. 1 Effect of particle size on amount of adsorption of Cu (II) (A) and chlorpyrifos (B) by cork waste at initial concentration: 10 mg/L and 1 µg/L, respectively; agitation time: 180 min.

Effect of contact time
The effect of contact time on the adsorption of Cu (II), Ni (II) and chlorpyrifos was investigated at different time intervals in the range of 0 to 300 min at 9.5, 8.75 mg/L and 1 µg/L, respectively. The results were presented for three pH in the case of heavy metals (5, 6 and 7) and three particle size (2-3, 3-7 and 20-30 mm) in the case of chlorpyrifos.

Cork Science and its Applications (CSA2017)   Materials Research Forum LLC
Materials Research Proceedings 3 (2017) 75-83   doi: http://dx.doi.org/10.21741/9781945291418-9

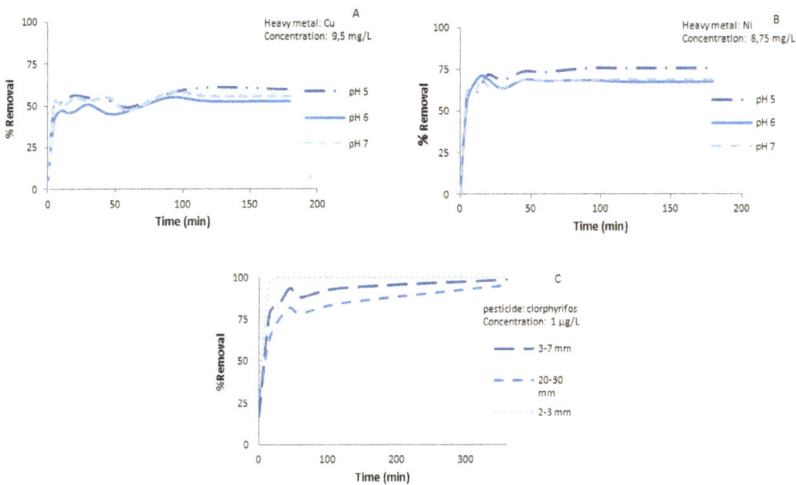

Fig. 2 Effect of contact time on amount of adsorption of Cu (II) (A), Ni (II) (B) and chlorpyrifos (C) by cork waste at initial concentration: 9.5, 8.75 mg/L and 1 µg/L, respectively.

Fig 2A and Fig 2B shows the effect of contact time amb pH on adsorption of Cu (II) and Ni (II) using cork waste. The results showed that the percentage of metal ion adsorption increase with increasing time and equilibrium was reached at the plateau value at 40 min. Thus, it has been chosen 180 min as equilibrium time to ensure the equilibrium in all concentration and experimental conditions. Due to the pH dependence of metal sorption, the sorption of both metal ions was studied at three different pH values (5, 6 and 7). The maximum removal (about 62%) for Cu (II) were found a pH 5 and differences on removal percentage were not found for Ni (II) at the different pH studied (from 65.3 to 68.6%). After the contact with cork waste, metallic solutions had a equal final pH of 6.5 independently the value of initial pH, indicating a buffering behavior of cork. The study continued at initial pH within the range of 5 to 6.

The effect of contact time on the adsorption of chlorpyrifos using cork waste at different granulometries was investigated and the results are presented in Fig 2C. The chlorpyrifos was totally sorbed rapidly within 30 min in particle size 2-3 mm (100% of recovery). Equilibrium time was reached within 120 min in higher particles size. As in the case of heavy metals, for pesticides it has been chosen 180 min as experimental time to ensure the equilibrium in all experimental conditions.

Effect of the amount of organic material
In order to take into account the adsorption conditions that could occur during the treatment of water in a wetland, the presence of organic material using COD value was tested. Table 1 shows the differences regarding the value of COD in the three types of synthetic wastewater.

Fig 3. shows the effect of chemical oxygen demand on the adsorption of Cu (II), Ni (II), aldrin, chlorpyrifos and tebuconazole onto cork. There was a reduction in the adsorption of Cu (II) and Ni (II) onto cork as the COD value increases (598 mg/L or composition 3). The removal percentage of Ni (II) decreases from 43 to 32% and for Cu (II) decreases from 54.2 to 31.8% among composition 1 to 3 (Figure 3A). A slight increase in COD value results in significant decreases on metal ions removal. Thus, the presence of high concentration of COD would have a negative effect in metal elimination on the wetlands.

COD studied values did not affect the ability of cork waste as adsorbent of targeted pesticides. The percentage of removal was from 98.6 to 100% for aldrin, from 39 to 54.5% for tebuconazole and 100% for chlorpyrifos for composition 1, 2 and 3 (Figure 3B). It seems that the presence of organic material do not affect pesticides removal on wetlands.

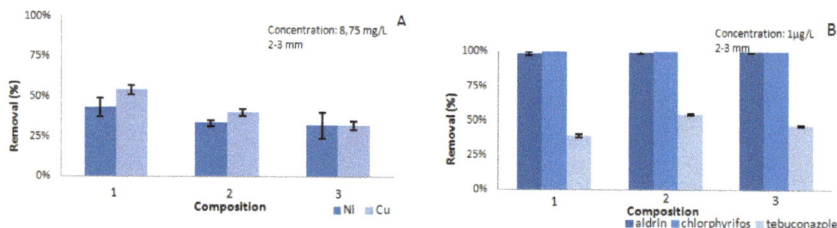

Fig. 3. Effect of organic material on amount of adsorption of Cu (II) (A), Ni (II) (B) and aldrin, chlorpyrifos and tebuconazole (B) by cork waste at initial concentration: 8.75 mg/L and 1 μg/L, respectively; agitation time: 180 min.

Effect of temperature
As in the case of COD, temperature was taken into account to evaluate the effect of the wetland conditions in adsorption. Two temperatures, 5 and 20°C was chosen, simulating winter and summer climatic conditions. As seen in Fig. 4, removal percentages of heavy metals (Fig 4A) and pesticides (Fig 4B) were maintained regardless of the temperature.

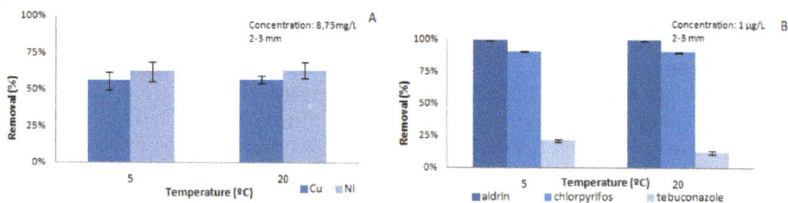

Fig. 4 Effect of temperature on amount of adsorption of Cu (II) (A), Ni (II) (B) and aldrin, chlorpyrifos and tebuconazole (B) by cork waste at initial concentration: 8.75 mg/L and 1 μg/L, respectively; agitation time: 180 min.

Temperature value did not affect the ability of cork waste as adsorbent of heavy metals and pesticides.

Cork waste as a constructed wetland substrate for pesticides and heavy metal removal
Three constructed wetland, two horizontal (HW1, HW2) and one vertical (VH), filled with cork waste were monitored during 16 days. Table 2 presents the removal percentages of each target pollutant for each type of wetland after 16 days of operation.

Table 2. Removal percentage of pesticides and heavy metals in each type of wetland

	% REMOVAL (16 days)		
	HW 1	HW 2 control	VW
Metalaxyl	99,6		99,6
Chlorpyrifos	100,0		100,0
Tebuconazole	99,4		99,2
Cu (II)	98,1	85,4	98,0
Ni (II)	98,1	71,9	97,5

The results showed that cork as a filled of horizontal and vertical wetland presents percentages of pollutant removal from 71.9% to 100%. The percentage removals were high for both pesticides and heavy metals. The horizontal wetland (HW1) (from 98.1 to 100 %) seems to be a bit more efficient than the vertical wetland (from 97.5 to 100%).

The initial concentration of heavy metal could affect the removal capacity of cork waste considering the results obtained for the control wetland (HW2). HW2 was only fed from winery wastewater without fortification. In this case, it was observed that the initial water had already traces of Cu (II) and Ni (II) (7.2 µg/L and 4.3 µg/L, respectively) and that were eliminated by filled cork (85.4 and 71.9%, respectively).

The concentration of pesticides and heavy metal in wetlands effluents throughout the study showed in Fig 5 and Fig 6.

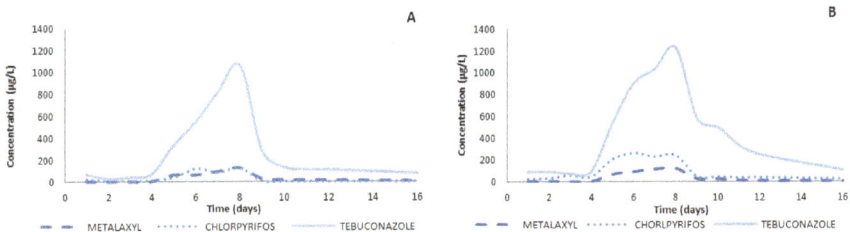

Fig. 5 Concentration of metalaxyl, chlorpyrifos and tebuconazole of wetland water effuent, HW1 (A) and VW (B), throughout its operation.

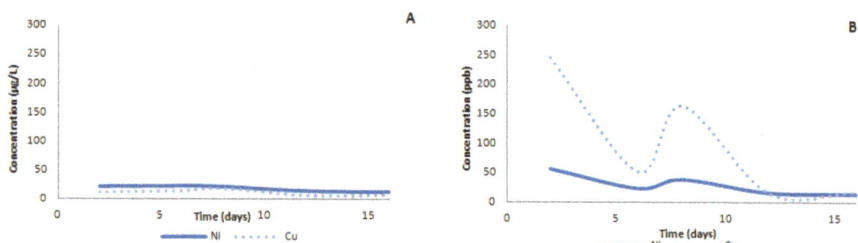

Fig. 6 Concentration of Cu (II) and Ni (II) of wetland water effluent, HW1 (A) and VW (B), throughout its operation.

In the case of pesticides, the effluent concentrations increased during the contamination period and then decreased over time. The effluent concentrations of pesticides in control wetland (HW2) were lower to the limit of detection (data not shown). In both wetlands, concentrations of all pesticides reached the maximum values at day 8 coinciding with the last contamination. For HW1 (Fig 5A), the effluent concentrations of metalaxyl (from 122.0 to 0 µg/L) and chlorpyrifos (from 123.8 to 1.9 µg/L) were lower than tebuconazole (from 1057.6 to 23.9 µg/L). For VH (Fig 5B), also the concentration of metalaxyl (from 118.6 to 0 µg/L) and chlorpyrifos (from 239.5 to 20.3 µg/L) were lower than tebuconazole (from 1223.0 to 75.2 µg/L). In batch experiments, tebuconazole also presented the lowest removal percentages. According to these results, cork waste seems to be more efficient for the adsorption of chlorpyrifos and metalaxyl than for tebuconazole.

The evolution of the effluent concentration of Cu (II) and Ni (II) depends on the type of wetland. For HW1 (Fig 6A), heavy metal concentrations had remained constant throughout the period of contamination and decontamination. The ranges were from 13 to 22 µg/L in the case of Ni (II) and from 5.6 to 18 µg/L for Cu (II). For VH (Fig 6B), Cu (II) and Ni (II) initial concentrations were higher than HW1, then decreased to increase again and finally decreased. This behavior was more evident for Cu (II) than for Ni (II). The ranges were from 14 to 56 µg/L in the case of Ni (II) and from 15 to 245 µg/L for Cu (II).

Conclusions
- ✓ Cork waste particle size affects the adsorption efficiency of targeted pesticides and heavy metal adsorption. The best tested size was 2 -3 mm.
- ✓ The high efficiency of cork waste as a sorbent of these pollutants was obtained after 40 minutes of contact.
- ✓ The presence of organic matter leads to a decrease in the quality of metal ions adsorbed and did not affect the pesticides removal.
- ✓ Temperature did not affect the removal percentage of pesticides and heavy metal studied.
- ✓ In the case of pesticides, cork waste was more adsorbent for chlorpyrifos and aldrin than for tebuconazole.

✓ Cork waste as a substrate of horizontal and vertical wetland allowed the removal of tebuconazole, metalaxyl, chlorpyrifos, Cu (II) and Ni (II) present in polluted winery wastewater.

✓ The horizontal wetland seemed to be a bit more efficient than vertical wetland for the removal of selected pollutants.

Acknowledgments
This study has been carried out within the framework of the LIFE project ECORKWASTE, LIFE14 ENV/ES/000460.

References

[1] Cabras, P., Angioni, A. Pesticide residues in grapes, wine and their processing products. J. Agric. Food and Chem., (2000), 48 (4) 967-973. https://doi.org/10.1021/jf990727a

[2] Chubar, N., Carvalho, J.R., Correia, M.J.N. Cork biomass as biosorbent for Cu(II), Zn(II) and Ni(II). Colloid. Surf. A: Physicochem. Eng. Asp. (2003), 230, 57-65. https://doi.org/10.1016/j.colsurfa.2003.09.014

[3] Chubar, N., Carvalho, J.R., Correia, M.J.N. Heavy metals biosorption on cork biomass: effect of the pre-treatment (2004). Colloid. Surf. A: Physicochem. Eng. Asp. (2003) 2, 38 (1-3), 51-58. https://doi.org/10.1016/j.colsurfa.2004.01.039

[4] Domingues, V. Utilizacão de um produto natural (cortiça) como adsorvente de pesticidas piretróides em águas. Faculty of Engineering - University of Porto, Porto, 2005.

[5] Domingues, V.F., Priolo, G., Alves, A.C., Cabral, M.F., Delerue-Matos, C. Adsorption behavior of alpha-cypermethrin on cork and activated carbon. J. Environ. Sci. Health., Part. B., (2007),42 (6) 649-654. https://doi.org/10.1080/03601230701465635

[6] Hanzlík, J., Jehlicka, J., Sebek, O., Weishauptová, Z., Machovic, V. Multicomponent adsorption of Ag(I), Cd(II) and Cu(II) by natural carbonaceous materials. Water Res., (2004) 38 (8), 2178-2184. https://doi.org/10.1016/j.watres.2004.01.037

[7] Park, D., Yun, Y.S. Park, J.M. The past, present and future trends of biosorption. Biotech.Bioprocess Eng., 15 (2010) 86-102. https://doi.org/10.1007/s12257-009-0199-4

[8] Villaescusa, I., Fiol, N., Cristiani, F., Floris, C., Lai, S., Nurchi, V.M. Copper(II) and nickel(II) uptake from aqueous solutions by cork wastes: a NMR and potentiometric study. Polyhedron., 21 (2002) (14-15) 1363-1367.

[9] Villaescusa, I., Martínez, M., Miralles, N. Heavy metal uptake from aqueous solution by cork and yohimbe bark wastes. J. Chem. Technol. Biotechnol. (2000) 75 (9), 812-816. https://doi.org/10.1002/1097-4660(200009)75:9%3C812::AID-JCTB284%3E3.0.CO;2-B

[10] Volesky, B. Detoxification of metal-bearing effluents: biosorption for the next century Hydrometallurgy, 59 (2001) 203–216. https://doi.org/10.1016/S0304-386X(00)00160-2

Keyword Index

About the Editors

Ricardo Alves de Sousa (São Paulo, 7th October, 1977) is currently Assistant Professor at the Department of Mechanical Engineering and member of the Center of Mechanical Technology (TEMA) research unit. In 2006, he obtained his PhD degree in Mechanical Engineering from the University of Aveiro, Portugal.

He has more than 70 scientific contributions either in papers, book chapters and books, and he is author of 3 patents. In 2011, he received the international scientific ESAFORM (European Association of Material Forming) career prize. In 2013 he received the Innovation Prize from APCOR (Portuguese Association for Cork), and in March of 2015 he became researcher of the month at the University of Aveiro.

Ofélia Anjos is a professor at the Agrarian School of Polytechnic Institute of Castelo Branco. She has obtained her BSc in Forest products at the University of Lisbon and her MSc degree in Science and Technology on paper and other forest products at the University of Beira Interior and University of Aveiro. She later obtained her PhD degree in Materials Science from the University of Lisbon (Instituto Superior Técnico), Portugal.

Ofélia Anjos is author of several papers in international peer reviewed journals, related to forest products, namely cork and more recent in food products (as example spirits and hive products) and infrared spectroscopic techniques.

www.ingramcontent.com/pod-product-compliance
Lightning Source LLC
Chambersburg PA
CBHW071500210326
41597CB00018B/2632